T0319966

Rational Behaviour and
the Design of Institutions

Rational Behaviour and the Design of Institutions

Concepts, Theories and Models

Hannu Nurmi
Professor of Political Science, University of Turku, Finland

Edward Elgar
Cheltenham, UK • Northampton, MA, USA

Published by
Edward Elgar Publishing Limited
The Lypiatts
15 Lansdown Road
Cheltenham
Glos GL50 2JA
UK

Edward Elgar Publishing, Inc.
William Pratt House
9 Dewey Court
Northampton
Massachusetts 01060
USA

This book has been printed on demand to keep the title in print.

A catalogue record for this book
is available from the British Library

Library of Congress Cataloguing in Publication Data
Nurmi, Hannu.
    Rational behaviour and the design of institutions: concepts,
theories, and models / Hannu Nurmi.
    Includes bibliographical references and index.
    1. Political science—Decision making. 2. Rational choice theory.
    1. Title.
JA71.N77 1998
302.3—dc21

98–12818
CIP

ISBN 978 1 85898 804 7

Printed and bound in Great Britain by
Marston Book Services Limited, Didcot

Contents

Figures

Tables

Preface

The concept of institution is somewhat elusive. Sometimes the word is used as a synonym for a set of norms (for example parliamentary system); sometimes it denotes a behavioural regularity of some sort (for example two-party system). Some institutions have a status (for example parliament); some are foci of authority and even worship (for example churches). This book discusses tools that can be utilized in the design of institutions. Our primary focus is on political institutions that produce outcomes (collective choices, policies) but most of the tools introduced are neutral as far as their application is concerned. In other words, they can be used in the design of any kind of institution.

Our approach to institutions is evaluative. It is based on the idea that institutions have a purpose with respect to which their functioning can be assessed. Sometimes finding out the purpose of an institution is difficult or controversial. Nevertheless, institutions are being born and modified all the time and there is considerable demand for information that could facilitate their design. Since our interest is in decision-making institutions and institutions that in general can be evaluated in terms of performance *vis-à-vis* some explicit criteria, it is only fair to admit that other approaches to institutions may complement the picture that our tools are able to draw.

We are dealing with an age-old problem. The question of how to guarantee through institutional means the preconditions for good life in a state has already been hotly debated in the ancient world. Although it was — and occasionally still is — thought that great individuals could overstep the institutional boundaries and thereby bring glory to themselves, institutions are generally viewed as a source of order and, consequently, of good life. This book deals with conceptual approaches, theories and models that shed light on how various social and political states emerge as outcomes of interactions among the goal-directed rational behaviours of individuals. As its title suggests, the book is intended as an exploration of the links between rational behaviour and institutional design. It provides tools for evaluating existing institutions and for setting up new ones.

This book is based on the lectures that I have given for advanced undergraduate and graduate students over a number of years. Many students have contributed to the present book with their comments on the lecture notes. I am very grateful to them. I am also most grateful to my friends and colleagues at the Political Science Department at the University of Turku: Kaisa Herne, Antti Pajala, Maija Setälä and Matti Wiberg, who have read parts of this book and given me their detailed comments. Tommi Meskanen has written the programs for computing the power index values and has helped me in typesetting the text. The numerous technical suggestions by Fiona Peacock of Edward Elgar are gratefully acknowledged. I would also like to thank Steven J. Brams and Keith Dowding for support and encouragement in this project. The work has been financially supported by the Academy of Finland. I dedicate this book to Dante and Malla, two splendid companions.

1 Introduction

This book deals with some important tools for understanding politics. We shall introduce and evaluate various theoretical constructs, for example concepts, models and theories, with respect to their ability to enlighten us about politics. Our discussion will be restricted to constructs based on the view of politics as rational action. This restriction is significant. Much of past and present theorizing about politics is based on different views. The reason for the restriction is simply that very often, if not always, when we encounter a social situation calling for an explanation − be it a strange kind of behaviour, a puzzling feature in some organization, an out-of-place utterance by a person or an institution that is somehow unique − we are satisfied with an account that tells us that the situation was after all an outcome of the rational actions of individuals or groups involved. In other words, explanations invoking rational actions are often at least intuitively satisfactory. They enlighten us about the reasons underlying behaviour or behaviour expectations. The latter, in particular, are important in understanding patterns of behaviour, political structures and institutions.

Rational action theories can be used in the pursuit of the traditionally recognized goals of scientific research, namely in explanation and prediction. In explaining events, behaviour, occurrences, and so on we give answers to 'why?' questions, whereas in predictions the questions concern what will happen. But despite the fact that philosophers of the social sciences often focus on behaviour of individuals as the *explanandum* of social research, we are often interested in making sense of other kinds of things, for example of structures, institutions, patterns of behaviour and so on. If a construct could only help us in dealing with individual behaviour, its usefulness would be severely limited.

Theories of rational action can, however, be used in explaining the emergence and stability of institutions. In fact, it is for this activity that these theories are particularly well suited. Many institutions are intentionally designed with some goal-states in mind. Thus, a representation system is designed with more or less detailed outcomes in mind. In proportional systems the aim is a reasonable match of the seat distribution with support distribution. In first-past-the-post systems, on the other hand, the objective is to have one representative for each areal unit or constituency. Both types of systems are based on intuitions about justice. They are rational ways to intuitively desirable ends. We can explain the emergence of those systems by the underlying views of justice and their continuation by the existence of coalitions of beneficiaries large enough to uphold the systems (and coalitions of disadvantaged ones small enough to make system change impossible).

1.1   MODELS AS TOOLS FOR UNDERSTANDING POLITICS

The constructs we shall mostly be discussing are called models. Although the concept of model has many different meanings (see for example Achinstein 1968) we shall use it in a relatively loose sense: a model is a simplified representation of the research object useful for specific purposes, for example explanation, prediction, institutional design, policy evaluation. The models thus understood are instruments of research. A specific model may be adequate for prediction, but not useful in explanation or understanding of phenomena in terms of familiar processes.

Since the models are considered to be simplifications of reality, it does not make much sense to ask if they are true. They aren't. Yet it is meaningful to ask what the reality would be like if the models were strictly true. Or, to phrase it slightly differently, how much of what we observe can be captured by the model? *Ceteris paribus*, the larger part of our research object the model captures, the better. The role of models can be illustrated as in Figure 1.1.

Game- and decision-theoretic models are usually regarded as mathematical. The particular advantage of mathematical models is their amenability to manipulation according to precisely defined rules. Their disadvantage, on the other hand, is that rather little of the social and political reality lends itself to mathematical description. In other words, we have useful models of a relatively limited range of phenomena. To classify game- and decision-theoretic models as mathematical constructs misses, however, the main point of those models, namely their strategic nature. In most models to be dealt with in this book, the mathematics is very elementary. Indeed, most often we shall be concerned with ordinal scale measurements.

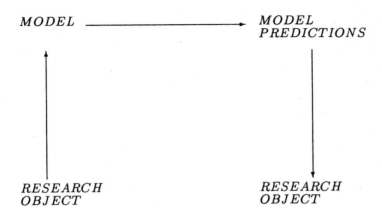

Figure 1.1: The Role of a Model

Although models are quite central in any factual science, there is practically no theory of model building. Indeed, this activity should perhaps be regarded as art rather than science. Despite this, in decision- and game-theoretical models there are some general guidelines as to what kind of entities the models should be composed of. The perspective in which the *primus motor* is an individual actor dictates that the available forms of action, the goals and aspirations of the actor as well as his/her[1] perceptions and anticipations ought to be taken into consideration in building models. In fact, these building blocks connect the theories to be discussed. What differentiates them is how the environment of the actor is described or modelled.

1.2 ORGANIZATION OF THE BOOK

The fundamental dividing line between the theories of rational action differentiates strategic environment theories from passive environment theories. Formerly this distinction coincided with the distinction between game and (individual and collective) decision theories. Today, the strategic view of environment has penetrated the collective decision theory as well. Thus, for example, it is common to analyse voting as a game

---

[1] Beginning in the next chapter we shall use the feminine gender in even-numbered and masculine gender in odd-numbered chapters for individuals, voters, players and so on.

of strategy.

This book will largely conform to the traditional classification of theories. Consequently, we shall start with individual decision theory since it deals with the basic concepts to be used in other theories as well, such as preference and utility. We shall discuss the axiomatization of rational behaviour as utility maximization. This view of rationality is admittedly controversial, but constitutes a suitable baseline model. We shall also discuss the empirical and normative relevance of rationality thus axiomatized.

Game theory is by now a very extensive field indeed. We shall restrict ourselves to the two-person games and, in particular, those of non-zero-sum variety. Of n-person games, those solution concepts with applicability to the constitutional design problem are touched upon.

The theory of collective decision making is also a very extensive field. Our focus is on voting paradoxes of various kinds as well as on the most important incompatibility results. With the aid of a decision- and game-theoretical conceptual apparatus we open a general perspective to the implementation of policies. Finally, we illustrate the institutional design problematique with approaches to the study of legislation in the European Union.

2 Rational Decisions

The most elementary setting discussed in the theory of rational behaviour is one in which there is a single decision maker faced with a choice problem in an essentially passive or disinterested environment. This setting is depicted in Figure 2.1.

Depending on how much she knows about the environment, the decision maker is operating under certainty, risk, or uncertainty. These three conditions are called decision modalities. Under certainty the decision maker knows basically all there is to know about her environment. She knows exactly what will happen when she makes any given choice. In effect, she is essentially choosing between consequences since everything leading to them is known. Under risk, on the other hand, the decision maker knows that her choices, together with the prevailing state of the environment, determine the outcomes and these, in turn, the consequences. However, she does not know which is the prevailing state. Rather, she knows the objective probability of each state. Thus, for each choice she can make, the decision maker knows the probability distribution of the outcomes and consequences. Under uncertainty, finally, the decision maker does not know the objective probabilities of the states of environments. This is very often the case when an entirely new decision situation emerges.

Decision theory deals with principles of rationality under these three modalities. Briefly stated, it views rationality as utility maximization. It is important to see, however, that this view is not an assumption in decision theory, but a conclusion which follows from specific behavioural assumptions. To criticize the view that rational behaviour coincides with utility-maximizing behaviour thus means that one or more of the behavioural assumptions is untenable. Before going into the critique, let us take a closer look at the utility maximization view.

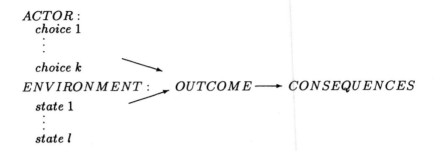

ACTOR :
  choice 1
    :
  choice k
ENVIRONMENT :　*OUTCOME* ⟶ *CONSEQUENCES*
  state 1
    :
  state l

Figure 2.1: The Setting of Decision Theory

2.1　UTILITY MAXIMIZATION UNDER CERTAINTY

The fundamental concept in decision theory is that of preference. The theory considers decisions as choices. Since the goal is to make sense of choice behaviour or make recommendations with regard to ways of making intelligent choices, it is essential that some standard of evaluation of the choices be available. In decision theory such a standard is provided by the decision maker herself. However, rather than evaluating the choices as such, the theory assumes that the decision maker has an opinion about the outcomes that ensue from the interaction of her choice and the state of the environment. This opinion is modelled as a preference relation. In behavioural terms the relation is captured in the following definition.

Definition 2.1 *The decision maker (strictly) prefers A to B, denoted by $A \succ B$, if she always (i.e. with probability 1) chooses A rather than B whenever she can choose between them.*

This definition establishes a conceptual link between preference and choice behaviour. The choices reveal preferences. The primitive (that is not defined) concept in the theory is that of preference. It is viewed as a binary relation between alternatives from which the choice is to be made. The preference relation can either be strict (denoted by '$\succ$') or weak (denoted by '$\succeq$'). The relation of indifference is denoted by '$\sim$'.

The intuitive content of the strict preference $A \succ B$ between outcomes A and B is that the former is regarded as better than the latter. The weak preference $A \succeq B$, in turn, means that A is considered to be at least as good as B, or that B is no better than A. The indifference between A and B, $A \sim B$, means that A is considered to be no better than B and B is considered to be no better than A.

The existence of a preference relation is, however, not enough to characterize rational behaviour. As a step towards this kind of characterization let us consider the following standard conditions, often called axioms, imposed on a preference relation.

Axiomatization 2.1 *Preference axioms:*

(a) Transitivity of strict preference: Let $A \succ B$ and $B \succ C$. Then also $A \succ C$.

(b) Transitivity of indifference: Let $A \sim B$ and $B \sim C$. Then also $A \sim C$.

(c) Trichotomy principle: For all A, B: either $A \succ B$ or $A \sim B$ or $B \succ A$.

Now, the strict preference relation can be defined by the weak preference relation and negation as follows: $A \succ B$ iff both $A \succeq B$ and not: $B \succeq A$. (The abbreviation 'iff' stands for 'if and only if'). Similarly, the indifference relation can be defined as: $A \sim B$ iff both $A \succeq B$ and $B \succeq A$.

With the aid of the weak preference relation (a) − (c) can be stated as: $\succeq$ is a complete preordering over certain prospects. A binary relation is a complete preordering iff it is both connected (or complete) and transitive. A relation $\succeq$ is complete in set X of alternatives iff for any pair $x, y \in X$ either $x \succeq y$ or $y \succeq x$ (or both). It is transitive iff for any triple $x, y, z \in X$ if $x \succeq y$ and $y \succeq z$ implies that $x \succeq z$.

In order to characterize rational behaviour we need some further definitions.

Definition 2.2 $I(A) = \{B \mid A \succeq B\}$, *the set of those Bs to which A is weakly preferred.*

Similarly,

Definition 2.3 $S(A) = \{C \mid C \succeq A\}$, *the set of those Cs that are weakly preferred to A.*

Thus, $I(A)$ is the set of alternatives that are inferior to A in the sense that A is weakly preferred to them. Similarly, $S(A)$ is the set of alternatives that are better than or as good as A.

So far we have imposed two conditions, transitivity and completeness, on binary weak preference relations. On the basis of these conditions it is possible to order a set of alternative outcomes in a sequence starting from the best ones and ending in the worst ones (according to the underlying weak preference relation) so that each outcome appears only in one position in the sequence. It is possible, however, that there are several 'tied' outcomes in the same position.

The next step in our characterization of rational behaviour is the solution of a representation problem. The problem consists in finding the conditions under which a given property, in this case that of preference, can be measured. Since the primitive notion is the binary preference relation we are looking for the conditions that it has to satisfy to make possible the measurement of the desirability of the outcomes. The measurement simply means the assignment of numbers to research objects in a way that preserves the relations that one wants to measure. Thus the measurement of preference means that one fixes a set of conditions on the preference relations. Those conditions have to be such that they guarantee the existence of a way to assign numbers to outcomes so that those numbers represent the preference relations, that is preserve the relevant characteristics of pairs of outcomes. If successful, the search for those conditions gives us a guarantee that there exists a measurable property of outcomes that represents the preference relation. The results of measurements performed on outcomes, that is a set of numbers, thus represents the preference relations. Let us now look at those conditions on preference relations that guarantee measurability.

Definition 2.4 *Let U be a real-valued function over X (the set of certain prospects or alternatives) so that,*

$$U(A) \geq U(B), \;\; \text{iff } A \succeq B \text{ according to } i\text{'s preferences.}$$

Let moreover

$$U(A) > U(B), \;\; \text{iff } A \succ B$$

and

$$U(A) = U(B), \;\; \text{iff } A \sim B.$$

Then U is a utility function that represents i's preferences.

U represents the preferences by giving higher values to more preferred alternatives and equal values to alternatives between which the decision maker is indifferent. The choice behaviour of a decision maker that chooses according to her preferences looks now as if she would try to maximize U with her choices. It is worth emphasizing that the utility maximization cannot on the basis of these definitions be viewed as a cause or reason for choice behaviour. Rather, preferences give the underlying reason for choice and, given the additional restrictions, also allow for the existence of a utility function. The decision maker may or may not be aware of the existence of her utility function. If rational choices are dictated by preferences as the definition above suggests, then it is analytically true that rationality means utility maximization. In other words, if one wants to reject the view of rationality which states that all rational behaviour aims at utility maximization, then one has to reject either the assumption that preferences guide choices or the technical assumptions pertaining to the nature of the preference relation. These are stated as axioms in the following.

Axiomatization 2.2 *Axioms of rational behaviour under certainty:*

1. *The weak preference relation* $\succeq$ *is a complete preordering over* X.
2. *Continuity: for all* $A \in X$: $I(A)$ *and* $S(A)$ *are closed sets.*

To clarify the concept of closedness of a set, let us consider an infinite set of points, say, real numbers between 7 and 15. Consider any sequence of numbers in this interval that converges to some value in the interval. Convergence to value x means that no matter how small a number e one picks, say $e = 10^{-10}$, there exists an integer n so that after the nth element in the sequence, the difference between x and all remaining elements in the sequence is less than e. Consider now any such sequence $x_i, (i = 1, \ldots)$ that converges to x and assume that all x_i in the sequence are elements of a set S. The set S is closed iff for all sequences in which the members belong to S the convergence points (that is x) also belong to S.

From these axioms follows the representation theorem for decision making under certainty. A representation theorem is a solution to a representation problem. It is a result which states under which conditions the measurement of properties or events is possible. For example, in the measurement of probability the representation theorem deals with a binary relation of qualitative probability ('being at least as probable as') and states the requirements that this relation has to satisfy in order for a quantitative probability measurement to be meaningful. Similarly, in utility theory the representation theorem specifies the conditions which the preference relation has to satisfy to guarantee meaningful utility value assignment to alternatives.

Theorem 2.1 *Let the decision maker's preferences over certain prospects satisfy axioms 1 and 2. Then there exists a utility function U representing the decision maker's preferences over X and the behaviour of the decision maker is as if she would try to maximize U.*

The proof of this theorem can be found, for example, in Debreu's book (1959). The theorem only states the conditions that make a meaningful value assignment possible. There may be several ways of assigning utility values to alternatives, each satisfying the conditions mentioned in the theorem. Solving the representation problem by proving a representation theorem does not, however, solve another problem of measurement, namely determining the uniqueness of measurement. The solution of this problem tells us how unique is the way of assigning utility values to alternatives. Stated in another way, the solution indicates the scale of measurement. The following remark answers this question.

Remark 2.1 *Any U satisfying the above axioms is ordinal, that is unique up to monotone order-preserving transformations. In other words, if U is a utility function, so is V,*

$$iff\ V(A) = F(U(A)),\ for\ all\ A \in X,$$

where F is monotone increasing.

This remark thus states that the representation problem allows for infinitely many solutions, each satisfying the conditions of the representation theorem. If one set of assigning numbers to outcomes is found, any such way that preserves the order of the numbers assigned to outcomes will do as well. Thus, the numbers have but ordinal significance. No arithmetic operations to the measurement results lend themselves to meaningful interpretations.

Further reading: Harsanyi (1977); Roberts (1979).

2.2 UTILITY MAXIMIZATION UNDER RISK

When the decision modality is risk instead of certainty, the decision maker does not know the prevailing state of her environment. In techical jargon, she does not know the state of nature which, together with her choice, determines the outcome. In other words, she is only able to assign to each choice at her disposal a probability distribution over the ensuing outcomes. The decision maker is facing several risky prospects. A risky prospect is a probability distribution over certain outcomes. In deciding her choice the decision maker has to evaluate the choices in ways that are not solely dependent on the value of any single outcome. One particular way to do this is to utilize a utility function with the expected utility (EU, for brevity) property.

In accordance with commonly adopted notation, risky prospect C which results in outcome or prize A_1 with probability p_1, in outcome or prize A_2 with probability $p_2, \ldots$, and in outcome or prize A_k with probability p_k will be denoted $C = (A_1, p_1; \ldots; A_k, p_k)$. Outcomes which occur with zero probability along with the respective probabilities are simply left out. Thus, any certain outcome A can be regarded as a risky prospect $A = (A, 1; B, 0)$.

Definition 2.5 *A utility function has the EU property if it assigns to the risky prospect $C = (A, p; B, 1 - p)$ the utility value*

$$U(C) = U(A, p; B, 1 - p) = p \cdot U(A) + (1 - p) \cdot U(B).$$

The utility functions with this property are von Neumann–Morgenstern utility functions.

Thus, utility functions with the EU property determine the utility value of a risky prospect as a weighted average of the utility values of the outcomes so that the weights equal the probabilities of the outcomes. Obviously not all utility functions have this property.

We shall need two additional technical definitions. The first one defines the set of those risky prospects or probability distributions over two certain outcomes (A and C) that are no better than a given third certain outcome (B). The second, in turn, defines the set of those probability distributions involving two certain outcomes that are at least as good as a given certain outcome.

Definition 2.6 $I^* = I^*(B; \ A, C) = \{p \mid B \succeq (A, p; \ C, 1 - p)\}$.

Definition 2.7 $S^* = S^*(B; \ A, C) = \{q \mid (A, q; C, 1 - q) \succeq B\}$.

We now proceed to state the axioms or conditions used in the representation theorem for utility measurement under risk.

Axiomatization 2.3 *Axioms of rational behaviour under risk:*

1. *Complete preordering. The weak preference relation over Y (the set of risky prospects) is a complete preordering.*
2. *Continuity. For all A, B, C: $S^*(B; A, C)$ and $I^*(B; A, C)$ are closed sets.*
3. *Monotonicity in prizes. Let $A* \succ A$ and $p > 0$. Then $(A*, p; \ B, 1 - p) \succ (A, p; B, 1 - p)$, and conversely.*

The first two axioms are analogous to those used in the context of certainty modality. Indeed, if one were just interested in finding conditions under which a utility function exists, the third axiom would not be needed. However, it is needed since we are looking for conditions under which rational behaviour can be represented as expected utility maximization.

The content of the continuity axiom is somewhat different in the context of risky prospects from that under certainty. Consider two certain outcomes A and C so that the decision maker strictly prefers the former to the latter. Then the continuity axiom states that for any outcome B which the decision maker regards as worse than A but better than C, there is a risky prospect D or a probability mixture of A and C, namely

$$D = (A, p*; C, 1 - p*)$$

so that the decision maker is indifferent between D and B. In other words, the value of a risky prospect depends continuously on the probability values expressed in its definition. Thus, by adjusting the probability values $p*$ in the above expression, one can find, for any outcome, a probability mixture of two other outcomes so that the decision maker is indifferent between the outcome and the probability mixture.

The monotonicity in prizes axiom is also known as the independence axiom. It is an intuitively most plausible requirement which can be exemplified by considering two lotteries in which only one prize is to be won, the winning probability being the same in both cases. The axiom requires that one should prefer the lottery in which the prize is higher to that in which it is lower. The representation theorem can now be stated.

Theorem 2.2 *Let the decision maker's preferences satisfy complete preordering, continuity and monotonicity in prizes axioms for risky prospects. Then there exists a utility function U such that U represents the decision maker's preferences and has the EU property. The decision maker behaves as if she tries to maximize the expected value of U.*

The proof of this theorem can be found in Harsanyi's book (1977). This solution to the representation problem leaves open, however, the uniqueness of the measurement. The following remark gives an answer to this problem.

Remark 2.2 *The U function of the preceding theorem is cardinal, unique up to order-preserving linear (or affine) transformations, that is if U is a utility function, so also is V, if $V(A) = a \cdot U(A) + b$, for all $A \in Y, a, b \in \mathbf{R}$ and $a > 0$.*

Here $\mathbf{R}$ denotes the set of real numbers.

Thus, there are again infinitely many ways of assigning utility values to risky prospects even when the axioms are satisfied. Regardless of the way the values are assigned (provided that the axioms are satisfied by the underlying preference relation), there is a property that remains invariant in all measurements, namely the ratio of differences of measurements of alternatives.

The question of how much of human behaviour under risk can be captured by EU maximization can, thus, be restated as the question of how much of the behaviour satisfies the above axioms. If the observed behaviour deviates from EU maximization, then either the decision maker is not choosing according to her preferences or her preferences do not satisfy all the axioms.

Further reading: Harsanyi (1977); Luce and Raiffa (1957).

2.3 RATIONAL BEHAVIOUR UNDER UNCERTAINTY

When the decision maker is acting under uncertainty she does not have objective probability distributions over the states of nature and, consequently, her choices cannot be regarded as risky prospects. Rather, they are uncertain prospects, that is the decision maker has an idea of which outcomes may ensue from her choices, but does not have objective probabilities in respect of them. She only knows what the outcomes are, provided that certain conditioning events occur. In other words, the decision maker knows, for example, that the outcome is A if event e occurs and B if e does not occur (or the complement of e occurs). This kind of uncertain prospect is denoted: $(A|e; B|\bar{e})$.

The axiomatic characterization of rational behaviour under uncertainty follows the pattern of the previous section except for the additional complication that in the absence of objective probabilities, there obviously cannot be utility functions with the EU property. Instead, one aims at finding conditions under which a utility function with a subjective expected utility (SEU, for brevity) property exists so that each uncertain prospect can be associated with a subjective expected utility value that represents the decision maker's preferences. Thus, given two uncertain prospects $C = (A|e; B|\bar{e})$ and $D = (E|e; F|\bar{e})$, we try to establish conditions under which a utility function U exists so that $C \succeq D$ iff

$$p^e \cdot U(A) + p^{\bar{e}} \cdot U(B) \geq p^e \cdot U(E) + p^{\bar{e}} \cdot U(F),$$

where p^e and $p^{\bar{e}}$ are independent of outcomes $A, \ldots, F$.

For axiomatic characterization of rational behaviour under uncertainty we need the following axioms (Harsanyi 1977). (See also Anscombe and Aumann 1963; Savage 1954).

Axiomatization 2.4 *Axioms of rational behaviour under uncertainty:*

1. $\succeq$ *is a complete preordering over uncertain prospects* Z.
2. *EU property for risky prospects.*
3. *Monotonicity in prizes of uncertain prospects. Let* $A* \succ A$. *Then*
   $(A * |e; \; B|\bar{e}) > (A|e; \; B|\bar{e})$.

Consider now a situation in which there is a maximal (best) outcome Q and a minimal (worst) outcome R. For any other outcome A we have: $Q \succeq A \succeq R$ and $Q \succ R$. Then there exists p_A such that

$$(Q, p_A; \; R, 1 - p_A) \sim A.$$

p_A is the characteristic probability of A. In other words, there exists a probability mixture of the best and worst outcomes that in the decision maker's view is equally desirable as A.

We now proceed to the representation theorem for uncertainty modality. Before that, however, we define the utility function with the SEU property.

Definition 2.8

$$p^e = U(Q|e; \; R|\bar{e})$$

$$p^{\bar{e}} = U(Q|\bar{e}; R|e),$$

where p^e *and* $p^{\bar{e}}$ *are the decision maker's subjective probabilities concerning the occurrence of* e *and the non-occurrence of* e. *A utility function has the SEU property if it assigns the uncertain prospect the utility*

$$U(A|e; B|\bar{e}) = p^e \cdot U(A) + p^{\bar{e}} \cdot U(B).$$

The representation theorem establishes the existence of a utility function with the SEU property.

Theorem 2.3 *Let* p^e, $p^{\bar{e}}$, Q *and* R *be defined as above. Then,*

$$p^e + p^{\bar{e}} = 1$$

and

$$U(A|e; \; B|\bar{e}) = p^e \cdot U(A) + p^{\bar{e}} \cdot U(B), \; \text{for all } C = (A|e; \; B|\bar{e}).$$

The subjective probabilities used in computing the utility of the uncertain prospect are defined in terms of utilities assigned to the prospect of getting the best outcome, given that e occurs, and worst outcome, given that e does not occur. The latter utility can, in turn, be expressed in terms of probability p' of some event that satisfies the following condition:

$$(p^e = U(Q|e; R|\bar{e}) = U(Q|f; R|\bar{f}) = U(Q, p'; R, 1 - p')$$

Thus, the objective probability of the risky prospect can be used in computing the utility value assigned to the uncertain prospect.

The above axioms together with the assumption that the decision maker chooses according to her preferences guarantee that her behaviour can be described as SEU maximization. The ensuing utility function is cardinal as in the case of von Neumann–Morgenstern utilities, that is unique up to linear transformations that preserve the order of prospects.

Further reading: Anscombe and Aumann (1963); Harsanyi (1977); Luce and Raiffa (1957); Savage (1954).

2.4 INTUITIVE PRINCIPLES OF RATIONALITY

In the discussion of the preceding section rationality is basically seen as behaviour which follows the preferences. In fact, the core idea of the previous axiomatizations is that rationality is precisely choosing preferred alternatives. Under certain restrictions regarding the structure of preferences this behaviour then turns out to be representable as utility maximizing. The intuitive notion of rationality is more elusive. In intuitive terms a rational actor could, for example, be defined as follows.

Definition 2.9 *A rational actor is an actor making optimal choices in specified environments.*

Optimality is usually understood as efficiency in goal-attainment. Usually one thinks of principles of choice that eliminate certain courses of action as irrational or suggest certain principles which enable the actor to rank-order choices. Thus, for example, the following principles could be used in characterizing what is rational choice behaviour.

Criteria for efficient or optimal choices:

(a) expected utility maximization (EU or SEU)
(b) dominance criterion
(c) maximin criterion

Criterion (a) dictates the choice of that alternative which maximizes the expected utility ensuing from the outcomes. In other words, (a) chooses that risky prospect which has the largest expected utility. Criterion (b), on the other hand, is based on the binary relation of dominance, denoted by D, which is defined over pairs of choices. Alternative a dominates alternative b, in symbols aDb, iff a leads to at least as good an outcome as b under each state of nature and to a strictly better outcome than b in at least one state. Clearly, relation D is not necessarily complete, that is it may well be that for some pair $x, y \in X$ neither xDy nor yDx. Criterion (b) can be given two plausible construals:

- weak dominance criterion: the choice has to be made from undominated alternatives
- strict dominance criterion: if an alternative dominates all other alternatives, it ought to be chosen.

Obviously, whenever an alternative satisfies the strict dominance criterion, it also satisfies the weak one, but not *vice versa*. Criterion (c) states that one should choose an alternative that maximizes the security level. The security level, in turn, means the minimum utility that may result from a choice.

These three principles are not strictly comparable since (b) and (c) do not require any kind of probability values, whereas (a) cannot be applied in their absence. It is worth observing, however, that when a strictly dominant alternative exists (that is an alternative satisfying the strict dominance criterion), it *eo ipso* satisfies criterion (c) as well. It is also clear that a strictly dominant alternative will be chosen under any probability distribution over states of nature. Thus it also satisfies criterion (a).

When a strictly dominant alternative does not exist, criteria (a) and (c) can lead to conflicting choices. However, both criteria are compatible with the weaker version of (b) in the sense that both (a) and (c) choose some alternative(s) from the set satisfying the weak dominance criterion.

What follows is a 'political' version of an example discussed by Chernoff and Moses (1957). It illustrates the preceding observations.

Example. In the late 1980s the countries in northern Europe had to take a stand on developments in the Baltic states and, in particular, on the rapidly growing demands for independence. At that time Estonia, Latvia and Lithuania were Soviet republics. With the imminent collapse of the Soviet Union, ever louder voices for independence had been raised both in the Baltic states and abroad. The governments of Denmark, Finland, Norway and Sweden had each somewhat differing views on the proper approach to these developments. In particular, from the Finnish viewpoint basically three courses of action could be envisioned:

Losses State	deny support to independence a_1	remain neutral a_2	give support to independence a_3
no collapse s_1	0	1	4
collapse s_2	5	3	2

Table 2.1: The Baltic States Policy Problem

- a_1: to consider the Baltic incidents as internal matters of the Soviet Union and refrain from expressing sympathy for the forces demanding independence
- a_2: to remain neutral on the issue of independence and to try to encourage negotations between the Soviet government and representatives of the forces striving for independence
- a_3: to give explicit support to the forces striving for independence.

The relevant states of nature in this decision problem were:

- s_1: the Soviet Union will not collapse in the foreseeable future
- s_2: the Soviet Union will soon collapse.

On the basis of forecasts regarding the level of welfare of the Finnish economy and society under these scenarios (each scenario being a combination of an action and a state of nature), the following table of losses is plausible in view of the public discussion (see Table 2.1).

The entries in the table indicate the decision maker's subjective evaluation of loss ensuing from her choice of policy, represented by the corresponding column, if the state of nature were that represented by the corresponding row. Thus, for example, 0 on row s_1 and column a_1 means that should the decision maker consider the Baltic issue as being an internal affair of the Soviet Union not to be interfered with by outsiders and should there be no collapse of the Soviet empire, her loss would be 0. If, on the other hand, policy a_1 were adopted and the state of nature were s_2, the loss in good-will, trade and political terms would be maximal, that is 5.

Suppose that the Finnish foreign policy experts unanimously suggest that the probability of s_1 is $3/4$ and that of s_2 is $1/4$. Criterion (a) now takes the form of expected loss minimization, since maximizing expected utility is obviously equivalent to minimizing expected loss. The expected loss related to a_1, denoted by $EL(a_1)$, equals $5/4$, while $EL(a_2) = 3/2$ and $EL(a_3) = 7/2$. Thus, with these *a priori* probabilities a_1 is dictated by (a).

If we look at the table of losses from the point of view of the dominance relation, we notice that no choice dominates the others. Thus, there is no strictly dominant choice. On the other hand, the set of un-dominated choices includes all options. The maximin criterion which is defined for utilities takes the form of minimax when the problem is defined in terms of losses. In other words, one should now minimize the maximal losses. The maximal loss related to a_1 is 5, that related to a_2 is 3 and that related to a_3 is 4. Thus, this criterion calls for a_2 to be chosen.

Each of the above principles of rationality leads to problems, either conceptual or empirical. Criterion (c) is very pessimistic in focusing attention on the worst outcomes in each choice. Criterion (b), in turn, is often useless either because its strict version specifies no choice at all or its weak version includes too many choices, as was the case in the above example. Also criterion (a) leads to problems which are of an empirical nature. In numerous experiments conducted over the past decades it has turned out that individuals seem to deviate systematically from the EU maximization principle at least in certain types of settings involving choice from risky prospects. We shall turn to some of these deviations as well as their purported explanations in the following sections.

We shall conclude this section with a tree summarizing the proper-ties and suitable contexts of various intuitive principles of rational choice behaviour (Figure 2.2). The figure is a simplified version of Giere's tree (Giere 1979). The first node M of the tree has three branches, one for each decision modality. In the case of certainty (C), the rational de-cision maker chooses the best alternative. In the case of risk (R), the EU criterion becomes applicable, although, in view of the experimental evidence to be discussed in the following sections, one may hesitate to regard this as the sole principle of rationality. In the case of uncertainty (UC) there may exist a dominant alternative. If it exists, then a ratio-nal decision maker should choose it. Otherwise, resort must be sought in various rules of thumb. The minimax rule calls for minimizing the maximum loss related to each alternative. Although in some contexts (for example in two-person constant-sum games) clearly plausible, it re-flects more pessimism than rationality. What Giere calls the gambler's rule is a mirror image of the minimax rule: it calls for the choice of that alternative which is associated with the possibility of getting the largest payoff or smallest loss. Laplace's rule, finally, is the application of the EU criterion to uncertainty with the assumption that each state of nature occurs with equal probability. Clearly the three last-mentioned principles are rules of thumb and not rationality principles as such.

Further reading: Chernoff and Moses (1957); Giere (1979).

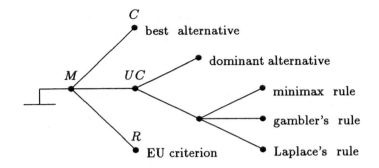

Figure 2.2: Intuitive Decision Principles

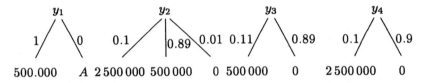

Figure 2.3: The Allais Paradox

2.5 THE ALLAIS PARADOX

As was stated above, the fact that behaviour deviates from utility maximization means either that the decision maker does not act according to her preferences or that at least one of the axioms of rational behaviour is not satisfied. Since the axioms are intuitively plausible, the observations contradicting utility maximization are often regarded as paradoxes. No doubt the best known of them is due to Allais, who in the early 1950s observed that the behaviour of experimental subjects under risk is not compatible with EU theory. Consider Allais's experiment.

The subjects were offered a choice from the following lotteries so that they could first choose either y_1 or y_2 and then either y_3 or y_4. Lottery y_1 is trivial: when choosing it one is certain to win $500\,000$ units of currency. Hence, any other outcome A occurs with probability 0. When choosing lottery y_2 the experimental subject can end up with one of three possible outcomes: either she wins $2\,500\,000$ or $500\,000$ or nothing. The probabilities of these outcomes are 0.1, 0.89 and 0.01, respectively. Lottery y_3 has two outcomes: $500\,000$ or nothing. Their probabilities are 0.11 and 0.89, respectively. Finally, lottery y_4 has the outcomes $2\,500\,000$ and nothing, with probabilities 0.1 and 0.9. The lotteries can thus be depicted as in Figure 2.3.

Allais found that her subjects tended to prefer y_1 to y_2, on the one hand, and y_4 to y_3, on the other. This behaviour is not only often observed but also intuitively plausible ('rational'). Yet, it contradicts rationality principle (a) and, consequently, EU theory. This becomes evident once we adopt the following notation.

Let $U(500\,000) = B_1$ and $U(2\,500\,000) = B_2$. Then the expected utilities of lotteries are:

$$
\begin{aligned}
U(y_1) &= B_1 \\
U(y_2) &= 0.1 \cdot B_2 + 0.89 \cdot B_1 \\
U(y_3) &= 0.11 \cdot B_1 \\
U(y_4) &= 0.1 \cdot B_2.
\end{aligned}
$$

Now according to (a) the fact that y_1 is preferred to y_2, together with the assumption that the decision maker is an EU maximizer, implies that $0.11B_1 > 0.1B_2$.

Similarly, the fact that y_4 is preferred to y_3, together with the EU maximization assumption, implies that $0.11B_1 < 0.1B_2$, that is, a contradiction. Thus, the decision maker cannot be an EU maximizer.

A few remarks ought to be made about the Allais paradox. Firstly, the observed preference of the majority of experimental subjects was clearly based on other considerations than the expected monetary return since the expected monetary value of y_1 is smaller than that of y_2. Thus, we could not observe the contradiction if the subjects were expected payoff maximizers. Secondly, given that the empirical preference relations are reliable — and there seem to be a fair number of replications of Allais's experiment to suggest that they are — the contradiction cannot be explained away by stating that people have different utilities of payoffs and that, since the EU theory is concerned with utilities rather than payoffs, the empirical observations do not pertain to relevant entities. This explanation does not work as the above EU value computations show. To end up with a contradiction between empirical preference relations and the EU theory, one does not need the assumption that the utilities of subjects over payoffs would be linear functions of the payoffs. In fact, one does not need to assume even that $B_2 > B_1$. All one needs to assume in order to derive a contradiction between the preferences of the majority of subjects and EU theory is that each monetary payoff has a fixed utility value.

Further reading: Allais (1979); Fishburn (1991); MacCrimmon and Larsson (1979).

	colour (and number) of balls		
	red	white or blue (60)	
options	(30)	white	blue
1	$100	$0	$0
2	$0	$0	$100
3	$100	$100	$0
4	$0	$100	$100

Table 2.2: Ellsberg's Paradox

2.6 ELLSBERG'S PARADOX

The Allais paradox casts a shadow over the descriptive value of the EU theory in risky environments. Similar doubts are raised by another paradox discovered by Ellsberg. Let us consider Ellsberg's example (see Table 2.2). The decision maker is faced with the following situation. She knows that an urn contains 30 red balls and 60 balls which are either white or blue. Exactly how many of those 60 balls are white or blue is unknown to her. One ball is randomly chosen from the urn. Before the outcome of the choice (the colour of the ball) is disclosed to the decision maker she is given two problems: choose either option 1 or option 2 and, then, choose either option 3 or option 4. The colour of the ball chosen at the outset and the decision maker's choice of options determine the payoffs. Table 2.2 indicates the payoffs to the decision maker ensuing from different options and outcomes.

Now, if the decision maker chooses option 1 over option 2, she must think that the (unknown) proportion of blue balls is less than 1/3. This assumption would thus dictate the choice of option 3 rather than option 4. Yet Ellsberg argues that many people would choose 1 over 2, but 4 over 3. The latter choice behaviour is clearly inconsistent with EU theory. To see this, let the probability of blue balls be q and that of white balls $2/3 - q$. Now for the EU maximizer the preference of option 1 over option 2 entails:

$$1/3 \cdot U(\$100) > q \cdot U(\$100).$$

On the other hand, the preference of option 4 over option 3 yields:

$$2/3 \cdot U(\$100) > 1/3 \cdot U(\$100) + (2/3 - q) \cdot U(\$100),$$

whereupon we get:

$$q \cdot U(\$100) > 1/3 \cdot U(\$100),$$

which contradicts the first inequality.

Although both Allais's and Ellsberg's paradoxes raise doubts about EU theory as a descriptive theory, they point to rather different aspects of decision making under risk. The former pertains to choice behaviour in settings where one option yields a certain outcome. The latter, in turn, deals with decision-making situations where not all relevant probabilities are known. One way of expressing the difference between the paradoxes is to say that Allais's paradox may occur in situations where there is some uncertainty about utilities, whereas Ellsberg's paradox involves uncertainty about probabilities. Some authors have coined the behaviour in the latter paradox as ambiguity aversion (see, for example, French and Xie 1994). The common crucial feature in both paradoxes, however, is that the subjects systematically exhibit behaviour that is incompatible with the EU theory.

Further reading: Ellsberg (1961); Schick (1984).

2.7   PROSPECT THEORY

The Allais paradox and related puzzling patterns of behaviour under risk have given rise to a rich experimental literature. Particularly noteworthy are the contributions of Kahneman and Tversky. They suggest a theory of choice that accounts for some of the behavioural peculiarities which are not compatible with the EU theory. The starting point of the theory — called prospect theory — is a classification of behavioural deviations from EU theory. The deviations are called effects. The first of these is already familiar.

2.7.1   Certainty effect

In one of their many experimental setups Kahneman and Tversky confronted some of the subjects with the following choice problem.

Problem: Choose between A and B

A:	2500 with probability 0.33	B:	2400 with certainty
	2400 with probability 0.66		
	0 with probability 0.01		
	18%		82%

The rest of the experimental population was given the following problem.

Problem: Choose between C and D

C: 2500 with probability 0.33 D: 2400 with probability 0.34
     0 with probability 0.67   0 with probability 0.66
          83%   17%

The percentages indicate the portion of each subpopulation that chose the corresponding alternative. Now, if we apply the EU theory and assume that $U(0) = 0$, we notice that the preference of B over A implies that

$$U(2400) > 0.33 \cdot U(2500) + 0.66 \cdot U(2400)$$

which amounts to $0.34 \cdot U(2400) > 0.33 \cdot U(2500)$. On the other hand, the preference of C over D implies:

$$0.34 \cdot U(2400) < 0.33 \cdot U(2500).$$

We thus end up with a contradiction. Therefore, the behaviour of the majority of subjects is not consonant with the EU theory.[1] We notice that the above problem is in fact an instance of the Allais paradox. In the following we shall adopt a somewhat simplified notation of risky prospects suggested by Kahneman and Tversky. To wit, we shall omit all the outcomes with 0 payoffs. Similarly, the probabilities of those outcomes are also omitted. Thus, the prospect $(A, p; B, 1-p)$ is reduced to (A, p) if $B = 0$.

The characteristic feature of the Allais paradox is the presence of one certain alternative. Let us consider another similar example.

Problem: choose between A and B.

        A: $(4000, 0.80)$ or B: (3000)
             20%   80%

Problem: choose between C and D.

        C: $(4000, 0.20)$ or D: $(3000, 0.25)$
             65%   35%

The preference for B over A implies: $U(3000) > 0.80 \cdot U(4000)$, while the preference of C over D implies:

$$0.20 \cdot U(4000) > 0.25 \cdot U(3000),$$

that is the former means that $U(3000)/U(4000) > 4/5$ and the latter contradicts this.

---

[1] In the usual experimental setup, each group makes only one choice. Hence, it is not strictly correct to say that the majority of the subjects has a preference that deviates from the EU theory. However, under the assumption that the groups are randomly selected, this inference can be made.

2.7.2   Reflection effect

Another systematic deviation from EU theory discovered by Kahneman and Tversky is the reflection effect. It is encountered in choice situations under risk where some alternatives involve negative payoffs, that is losses instead of gains. Consider the following two problems.

Problem

$$A: \quad (-4000, 0.80) \quad \text{or } B: \quad (-3000)$$
$$92\% \qquad\qquad\qquad 8\%$$

Problem

$$C: \quad (-4000, 0.20) \quad \text{or } D: \quad (-3000, 0.25)$$
$$42\% \qquad\qquad\qquad 58\%$$

Note that the alternatives are the same as in the last example of the previous section except that now the payoffs are negative. The preference of the majority of subjects between negative prospects seems to be a mirror image of the preference between positive prospects, that is, the reflection of prospects around 0 reverses the preference order. Risk aversion in the positive domain is accompanied by risk seeking in the negative domain. Note that the preference of A over B and of D over C also violates the EU theory:

$$A \succ B \Rightarrow 0.80 \cdot U(-4000) > U(-3000),$$

while

$$D \succ C \Rightarrow 0.25 \cdot U(-3000) > 0.20 \cdot U(-4000).$$

Thus

$$A \succ B \Rightarrow U(-4000)/U(-3000) < 5/4$$

and

$$D \succ C \Rightarrow U(-4000)/U(-3000) > 5/4.$$

2.7.3   Isolation effect

The reflection effect suggests the possibility of modifying the choice behaviour by varying the representation of risky prospects. Indeed, Kahneman and Tversky's experiments on so-called framing have demonstrated that the description of risky prospects can be a significant determinant of choice behaviour. The following problem illustrates this.

Problem: The subjects participate in a two-stage game. In the first stage there is a probability of 0.75 that the game ends and one gets nothing. With a probability of 0.25 one moves to the second stage where the following choice situation is encountered. One must choose between $(4000, 0.80)$ and (3000). The subjects are asked to indicate their choice before entering the first stage (that is, before knowing whether they will ever reach the second stage). In effect the choice is between $(4000, 0.20)$ and $(3000, 0.25)$ as in the last example of certainty effect. However, now over 78 per cent choose $(3000, 0.25)$ in contradiction with the choice behaviour of the majority of subjects in that problem. Clearly, the description of the prospects seems to make a difference. Consider another illustration.

Half of the experimental population is confronted with Problem 1 and the rest with Problem 2.

Problem 1: In addition to whatever you own, you are given 1000 units of currency. Now choose between

$$A:\ (1000, 0.50) \quad \text{and}\ B:\ (500)$$
$$16\% \phantom{(1000, 0.50) \quad \text{and}\ B:\ } 84\%$$

Problem 2: In addition to whatever you own, you have been given 2000 units of currency. Now choose between

$$C:\ (-1000, 0.50) \quad \text{and}\ D:\ (-500)$$
$$69\% \phantom{(-1000, 0.50) \quad \text{and}\ D:\ } 31\%$$

The majorities are pretty clear: $B \succ A$ and $C \succ D$. However, a closer look at the prospects reveals that $A = (2000, 0.50; 1000, 0.50) = C$ and $B = (1500) = D$. Thus, assuming that the population has been randomly divided into the groups, we may infer that the difference in choice behaviour is due to the description of risky prospects.

2.7.4 Solution

The experimental anomalies have convinced Kahneman and Tversky of the necessity to develop a descriptive theory of individual choice behaviour that would adequately take into account the systematic deviations from the EU theory as exemplified by the above effects. Their theory — the prospect theory — is a qualitative one in the sense that it leaves many parameters of choice to be empirically determined, but outlines some qualitative features of individual choices under risk. In a nutshell the theory can be represented as in Figure 2.4.

The crucial qualitative characteristics of choice behaviour according to prospect theory are:

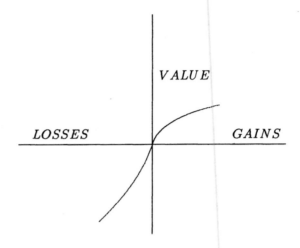

Figure 2.4: Prospect Theory

- risk-aversion on the positive domain
- risk-taking on the negative domain
- 'losses loom larger than corresponding gains'.

The first characteristic means that the individuals prefer a certain prospect to a risky one when the expected values of both prospects are the same and positive. The preference is reversed when the expected values are negative. The third characteristic means that the utility or value curves are steeper in the negative domain than in the positive one.

Prospect theory is able to accommodate the experimental anomalies discussed above. Indeed, the theory has been formulated specifically to do that. There are, however, puzzling behavioural invariances that deviate from EU theory and yet cannot be explained by prospect theory. One of them is preference reversal.

Further reading: Kahneman and Tversky (1979).

2.8   PREFERENCE REVERSAL

Another experimental anomaly concerning individual choice behaviour was discovered by Lichtenstein and Slovic in the early 1970s (Lichtenstein and Slovic 1971). It is more fundamental than the previous paradoxes because it calls into question the most basic notion of rational behaviour theory, namely the existence of a (weak) preference relation. Figure 2.5

depicts the experimental setup. The decision maker is asked to indicate her preference over two simple roulette games. In both games she is asked to spin a roulette wheel shown in Figure 2.5. Game A gives her $4 if the wheel stops with the pointer directed at any part of the circle except the vertical radius drawn inside the circle. If the pointer hits this radius, she will get nothing. In game B the decision maker gets $16 if the pointer hits the smaller sector within the circle. In B she gets nothing if the pointer hits some other spot in the circle.

Lichtenstein and Slovic made the psychological observation that A is preferred to B and yet the subjects were willing to pay more for the right to participate in B than to participate in A. This observation was later confirmed in a somewhat more carefully controlled experimental setup by Grether and Plott (1979). To be more precise, let

w = initial wealth of the decision maker;

$(z, 1, 0)$ = the outcome in which game A is chosen and wealth level is z;

$(z, 0, 1)$ = the outcome in which B is chosen and wealth level is z;

$(z, 0, 0)$ = the outcome in which neither A nor B are chosen and wealth level is z;

$\$(A)$ and $\$(B)$ are the respective selling prices.

It follows then, that:

1. $(w + \$(A), 0, 0) \sim (w, 1, 0)$ by definition of $\$(A)$
2. $(w + \$(B), 0, 0) \sim (w, 0, 1)$ by definition of $\$(A)$
3. $(w, 1, 0) > (w, 0, 1)$ because A is preferred over B
4. $(w + \$(A), 0, 0) > (w + \$(B), 0, 0)$ by transitivity
5. $\$(A) > \$(B)$ by positive value of wealth.

However, 5 is not observed in practice.

It is worth emphasizing that the anomaly here is not in the subjects' inability to indicate their preferences in respect of the games. Nor is it the case that they were indifferent between A and B. The puzzling feature is the apparent confusion between two reasonable operational definitions of weak preference. This phenomenon thus shows that the assumption according to which individuals have connected and transitive preferences is not so straightforward as one might think *prima facie*.

Further reading: Lichtenstein and Slovic (1971); Grether and Plott (1979).

Figure 2.5: Preference Reversal Experiment

2.9   REGRET THEORY

In everyday decision-making situations involving some uncertainty, one often feels either elation or regret once the outcome is known, depending on whether the action one has taken is in retrospect particularly appropriate or inappropriate with respect to one's goals. It seems possible that this feeling of elation or regret could play a role in individual decision making in general and perhaps suggests why certain types of deviations from EU theory occur. A theory based on this intuitive idea has been developed by Bell (1982), and Loomes and Sugden (1982).

An early predecessor of this theory is Savage's minimax regret theory (Savage 1954). Its motivation can be illustrated with reference to Table 2.1 on page 17. The entries in the table represent losses and reflect the welfare level of the decision maker *vis-à-vis* the present level. Thus, for example, the third number on the second row indicates that the decision maker would experience a 2-unit loss of welfare with respect to her present level if she chose to support the forces striving for independence and the Soviet Union did in fact collapse. Since the decision maker is not assumed to be able to affect the collapse, this may seem implausible: even though she makes the best possible decision in that state of nature, she still experiences a loss. That there would, in fact, be a loss in terms of short-term trading opportunities is not at issue here, but the fact is that losses seem to contain aspects that are beyond the decision maker's control and are thus irrelevant in the evaluation of success of the decision making. More relevant for the evaluation of the decision making itself would be a measure that reflects the amount of loss as a deviation from that ensuing from the optimal decision in the respective state of nature. This is the motivation underlying the concept of regret. In each state of nature the optimal decision is plausibly enough associated with zero regret, while non-optimal ones are associated with regrets computed as differences between their associated losses and that of the optimal one. In the Baltic states policy problem one thus obtains a table of regrets (Table 2.3).

On the basis of regrets Savage suggested a decision principle for situa-

Actions State	a_1	a_2	a_3
no collapse s_1	0	1	4
collapse s_2	3	1	0

Table 2.3: Table of Regrets in the Baltic States Policy Problem

tions involving uncertainty: the minimax regret principle. This principle amounts to choosing that alternative for which the maximum regret with respect to the unknown states of nature is minimal. In the policy problem, this principle would dictate the choice of a_2 since its maximum regret is 1, whereas the other alternatives have larger maxima.

It is clear that the minimax regret principle makes the choice dependent not only on the 'inherent' properties of alternatives but also on what other alternatives are available. The principle makes the choice behaviour in this sense context-dependent. In normative decision theory the possibility that an alternative, say a_i, is chosen from a large set A of alternatives and fails to be chosen from a proper subset of A using the same criterion, has been regarded as a serious shortcoming of the minimax regret principle (Chernoff 1954; Luce and Raiffa 1957). That the principle has this shortcoming can easily be seen from Table 2.4.

When all three actions are available, a_2 is chosen according to the minimax regret principle. However, if only a_1 and a_2 are available, a_1 should be chosen. Thus, the principle violates one type of consistency requirement, called property α, to be discussed in the next section.

Bell has suggested a modification of the regret concept that could explain, for example, why the same people gamble and buy insurance (Bell 1982). Gambling is perhaps the best example of risk-taking behaviour, while buying insurance aims at avoiding risks. In the case of gambling it is, moreover, to be observed that the expected value of most gambles is negative. Consider a situation where one can make a bet that a certain outcome occurs, say the next random draw from a deck of cards is the queen of spades. For the price p a bookie is willing to pay the gambler the amount of 1 if the card next drawn is, indeed, the queen of spades. Otherwise, the gambler gets nothing and the bookie keeps the price. This yields the outcomes shown in Table 2.5.

Suppose now that the gambler has the following kind of utility function over pairs of outcomes:

$$u(x_i, x_j) = v(x_i) + f(v(x_i) - v(x_j)).$$

where $v(x_i)$ is the value of x_i. In fact, $v(x_i) - v(x_j)$ is the regret of

Actions Environment	action a_1	action a_2	action a_3
state 1	0	2	4
state 2	3	2	0

Table 2.4: Context-Dependence of the Minimax Regret Principle

Actions Outcomes	bet	don't bet
queen of spades drawn	x_1	x_2
queen of spades not drawn	x_3	x_4

Table 2.5: The Outcomes of the Gambling Example

getting x_i rather than x_j. The function f thus reflects the importance of regret in the utility function.

If the alternative 'bet' is chosen and the outcome 'queen of spades' is observed, then the gambler's utility is $u(x_1, x_2)$. In a dichotomous choice situation like the one we are considering, it is natural to assume that an action is chosen rather than its complement if the expected utility of the former is larger than that of the latter. Suppose now that the gambler thinks that the probability that the next draw will result in queen of spades is p, so that the expected value of both betting and not betting is the same. We can predict that 'bet' is chosen instead of 'don't bet' iff

$$pu(x_1, x_2) + (1 - p)u(x_3, x_4) > pu(x_2, x_1) + (1 - p)u(x_4, x_3).$$

In addition to the table of outcomes, we can also construct the table of payoffs (Table 2.6).

In Bell's construal, the bet is the preferred action iff:

$$p[1-p+f(1-p)]+(1-p)[-p+f(-p)] > p[0+f(p-1)]+(1-p)[0+f(p)]$$

Without knowing f we cannot say much more about the choice behaviour regarding betting. However, it turns out that under plausible assumptions concerning insurance buying behaviour precisely those actors who would prefer betting to not betting would prefer getting insured to not getting insured against certain types of damages (for example car accidents). Consider the payoff table (Table 2.7).

Here the insurance premium is p and the loss in the case of an accident is 1. Assuming that the probability of the accident is p and resorting to Bell's calculus, getting insured is preferred to not getting insured iff

Actions Outcomes	bet	don't bet
queen of spades drawn	1−p	0
queen of spades not drawn	−p	0

Table 2.6: The Payoffs of Gambling Example

Actions Outcomes	get insured	don't get insured
accident occurs	−p	−1
no accident occurs	−p	0

Table 2.7: The Payoffs of Insurance Example

$$p[-p+f(1-p)]+(1-p)[-p+f(-p)] > p[-1+f(p-1)]+(1-p)[0+f(p)].$$

Comparing this with the gambling example reveals that the conditions under which getting insured is preferred to not getting insured are precisely those under which betting is preferred to not betting. Thus, gambling and buying insurance policies are not only mutually compatible, but can be seen as outcomes of identical principles of choice.

In the version of regret theory that Loomes and Sugden (1986) outline, the principle of choice is maximization of expected modified utility. Each available action a_i of the actor is viewed as an n-tuple of outcomes $x_i = (x_{i1}, \ldots, x_{in})$, one for each (unknown) state of nature. The probability distribution over the n states is known. Similarly the utility function $u(x_{ij}) = c_{ij}$ for all i and j is known. The actor is assumed to form an expectation $\bar{c}_i$ of the utility of each action a_i so that $\bar{c}_i = \Sigma_j p_j c_{ij}$. In other words, the expectation is simply the expected utility of action a_i. Once the action a_i has been taken and its utility consequence c_{ij} has unfolded, the individual experiences elation if $c_{ij} > \bar{c}_i$ and disappointment if $c_{ij} < \bar{c}_i$.

The effect of the elation or disappointment on the individual's choice behaviour is represented by a function D which maps the $c_{ij} - \bar{c}_i$ differences into utility increments or decrements. Thus the behaviour aims at maximizing

$$E_i = \sum_j p_j [c_{ij} + D(c_{ij} - \bar{c}_i)].$$

On the basis of this, one can construct a weak preference relation over actions as follows:

$$a_i \succ a_k \text{ iff } E_i > E_k$$

$$a_i \sim a_k \text{ iff } E_i = E_k.$$

This weak preference relation is obviously complete and transitive. If function D is linear, the choice behaviour based on maximization of E_i reduces to utility maximization. Thus, in order to be able to explain deviations from EU theory, one has to assume that D is non-linear. With this assumption, Loomes and Sugden are able to explain the various effects discussed in the prospect theory as simply following from the fact that the decision makers are maximizing modified expected utility. Of course, even though many types of behaviour can be explained by referring to the D function, there has to be an independent method of ascertaining for each decision maker the particular form of the function. Otherwise, one is vulnerable to the claim that the function works as an *ad hoc* device to make any behaviour explainable.

Further reading: Bell (1982); Loomes and Sugden (1986).

2.10   COMPARATIVE ADVANTAGE MODEL

The comparative advantage model is also designed to explain those behavioural deviations from EU theory that cannot be explained by prospect theory. Moreover, it is intended to accommodate also those behavioural invariances which are succesfully explained by prospect theory. In other words, the model purports to represent genuine scientific progress. Before describing the model, let us focus on two phenomena that cannot be accounted for by prospect theory.

2.10.1   Compromise effect

In political argumentation it is rather common to develop positions on an issue that could somehow be viewed as a compromise between the more extreme positions. Especially in situations where vote trading is not feasible, compromise positions often play a crucial role in decision making. What is called the compromise effect is a demonstration that describing a stand on an issue as a compromise can increase its desirability over what it would be otherwise. Let us look at an illustration (see Figure 2.6).

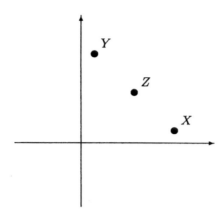

Figure 2.6: Compromise Effect

Suppose that policy alternatives X, Y and Z are being considered and that they can be represented as points in a two-dimensional space as indicated in Figure 2.6. Both dimensions can be viewed as relevant properties of policy alternatives — say the percentage of the population affected by the policy and the duration of the benefits of the policy — so that increasing positive values on both dimensions represent more desirable policy alternatives. Clearly, Z is better than X on the vertical dimension, while X is better than Z on the horizontal one. Now, there is some experimental evidence suggesting that Z's support relative to that of X increases when Y is present *vis-à-vis* its level when only X and Z are present. Thus, by introducing Y one would be able to increase the desirability of Z even though Y itself would have no chance at all of being chosen. This increase in Z's desirability would seem to follow from its status as a compromise between X and Y. Hence the label 'compromise effect'.

The effect points also to the violation of one of the basic postulates of EU theory, namely the context-free requirement. According to this requirement, the comparative evaluation of two options — here X and Z — is independent of which other alternatives are in the set of available options. The context-free requirement consists of two components, called property α and property β by Sen (1970). The former states that if an alternative a is chosen from a large set A of alternatives, then a should also be chosen from any proper subset of A. Property β, in turn, is a requirement that, should a and b be chosen from set A, then either both of them or neither of them should be chosen from any superset of A. Clearly the compromise effect, whenever present, demonstrates choice

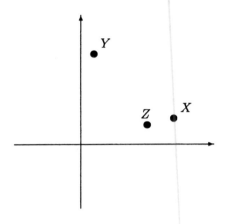

Figure 2.7: Asymmetric Domination

behaviour that violates either α or β.

The compromise effect was first found in research of consumer choice behaviour (Simonson 1989). One would expect it to surface also in political argumentation, whereby an extreme policy alternative is introduced into the debate not primarily as a realistic alternative to be adopted, but to make one's real policy favourite look more middle-of-the-road, that is, a compromise alternative. It is difficult to find hard evidence on the systematic use of the compromise effect in political argumentation, but it is not hard to find examples of debates where the parties have strived at not presenting their views as extreme ones, but rather as compromises (Herne 1997b). Sometimes this calls for proposing downright straw-man alternatives.

2.10.2   Asymmetric domination

Another effect of a similar genre as the compromise effect was also first studied in the consumer choice research (Huber et al. 1982; Shafir et al. 1990). Consider Figure 2.7 in which the dimensions and alternatives have the same interpretation as in the preceding example.

In the example alternative X dominates alternative Z in the sense of being better on both dimensions. Alternative Y, on the other hand, does not dominate Z. Hence, the term 'asymmetric domination'. In experimental studies it has been observed that the introduction of an asymmetrically dominated alternative increases the likelihood of the dominating alternative being chosen from what it had been if only two alternatives (neither dominating the other) had been available. In our example, the introduction of Z increases the likelihood of X being chosen over Y.

The asymmetric domination effect is intuitively plausible since it is based on the idea that alternatives 'compete' with each other. In this competition the decision maker makes pairwise comparisons. If an alternative is better than another in all dimensions, it is natural to think that the former defeats the latter. In our example, X thus defeats Z. Since one of the alternatives defeats another alternative, while none of the others does so, it makes intuitive sense to choose the only victorious alternative even though it does not defeat all the other alternatives. And yet, this intuition contradicts the context-free condition.

2.10.3 Explanation of effects

While the prospect theory helps to explain some types of choice behaviour that violate the axioms of the EU theory, it is still based on a similar underlying assumption concerning the nature of choice. To wit, both the EU theory and prospect theory assume that the main determinant of choice is the desirability of alternatives and the latter, in turn, is determined by the value that is assigned to alternatives in a context-free fashion. Since there is rather clear evidence that the choice is at least sometimes context-sensitive, as in compromise and asymmetric domination effects, one cannot be wholly satisfied with the account of choice behaviour provided by the EU and prospect theories. Moreover, the preference reversal phenomenon cannot be explained by these theories, either.

Shafir et al. (1989; 1990) have proposed a construct, the comparative advantage model, which aims at explaining the context-sensitive choice behaviour and preference reversal. Unfortunately, the model is based on a parameter which needs to be estimated independently of the choice setting for the model to be able to yield testable predictions. Let us outline the comparative advantage model.

Consider two lotteries:

$$L_1 = (A, p_1; 0, 1 - p_1)$$

$$L_2 = (B, p_2; 0, 1 - p_2)$$

where A and B are monetary payoffs so that $A > B$. Moreover, $p_2 > p_1$. Thus L_1 gives potentially a larger payoff than L_2, but the probability of winning is larger in the latter than in the former.

Let us call the expected payoffs $E(L_1)$ and $E(L_2)$ of the lotteries their absolute parts. Thus, $E(L_1) = p_1 A$ and $E(L_2) = p_2 B$. The comparative advantage of L_2 over L_1 or the probability advantage is simply $p_2 - p_1$. The comparative advantage of L_1 over L_2 or the payoff advantage, in turn, is $(A - B)/A$, that is, the improvement of A with respect to B as measured by units of A.

Now, the attractiveness of a lottery *vis-à-vis* another is determined by constant c_g which reflects the weight that the decision maker assigns to relative payoff differences versus probability differences. The choice behaviour, in turn, is determined by the relative attractiveness of lotteries. In other words, a lottery is chosen over another one just in case its attractiveness is larger than that of the other. L_2's attractiveness *vis-à-vis* L_1 is defined as

$$A(L_2) = E(L_2)(p_2 - p_1)$$

that is, the expected payoff multiplied by the probability advantage.
L_1's advantage over L_2, on the other hand, is defined as:

$$A(L_1) = E(L_1)\frac{A - B}{A} \cdot c_g.$$

According to the comparative advantage model, then, the comparative attractiveness of lotteries determines the choice. The subjective parameter c_g plays a crucial role in this calculus. Let us focus on an example in which the preference reversal phenomenon observed in experimental settings can be explained, provided that the c_g parameter has a certain value. Let L_3 and L_4 be defined as:

$$L_3 = (\$5, 0.80; 0, 0, 20)$$

$$L_4 = (\$10, 0.60; 0, 0.40).$$

In experiments it has been observed that L_3 is preferred to L_4 and, yet, the subjects are willing to pay more for L_4 than for L_3. Suppose that for a given subject $c_g = 0.25$, that is, the probability advantage weighs four times as much as the relative payoff difference in the attractiveness calculus of the subject. Obviously,

$$E(L_3) = 4, A(L_3) = 4 \cdot 0.20 = 0.80$$

$$E(L_4) = 6, A(L_4) = 6 \cdot \frac{5}{10} \cdot 0.25 = 0.75$$

Thus, L_3 is more attractive with respect to L_4 than the latter with respect to the former. Thus, one is able to explain why L_3 is chosen. But what about the higher value given to L_4?

Define now two other lotteries:

$$L_5 = (X, 1; 0, 0)$$

$$L_6 = (Y, 1; 0, 0)$$

so that L_5 is such a lottery that the decision maker is indifferent between L_3 and L_5. Similarly, L_6 is so designed that the decision maker is indifferent between it and L_4. We can obviously interpret X as the monetary value assigned by the decision maker to L_3 and Y as the monetary value assigned by her to L_4. Assuming, thus, that L_3 and L_5, on the one hand, and L_4 and L_6, on the other hand, are equally attractive, we get:

$$X(1 - 0.80) = 4 \cdot \frac{5 - X}{5} \cdot 0.25 \text{ or } X = 2.5.$$

Similarly, we get:

$$Y(1 - 0.60) = 6 \cdot \frac{10 - Y}{10} \cdot 0.25 \text{ or } Y = 2.7.$$

Thus L_4 has, indeed, a higher monetary value than L_3. It is worth emphasizing that not all reversals of preferences are explainable by the comparative advantage model. This would not be desirable, either, since it would deprive the theory of its empirical content and make it invulnerable to tests. However, for genuine testing one needs independent estimates of c_g.

Further reading: Shafir et al. (1989); Simonson (1989); Herne (1997a).

2.11 DYNAMIC CONSIDERATIONS

The bulk of the experimental literature on individual choice behaviour ends up with suggesting that in view of the important and widespread violations of the choice principles of EU theory, one should find alternative theories with more descriptive power. However, the observed violations do not as such provide a basis for the abandonment of the EU theory as a normative account. If people in general do not satisfy the EU principles, then it only means that they are not being rational, is the standard counterargument to the experimental findings. Indeed, the earliest justifications of the EU principles amount to showing that if people deviate from them, it is only at their own peril. So Friedman and Savage (1952) point out that a violation of EU principles, more precisely, the independence condition, amounts to violating the dominance criterion of choice, that is, the violator would wind up choosing a dominated alternative. Similarly, it has been shown that if the decision maker's choice behaviour deviates from the EU principle, a 'money pump' can be built so that the decision maker loses all her money by accepting a sequence of trades each one of which she considers profitable for herself

(Raiffa 1968). Suppose that a person's preferences over three alternative gambles g_1, g_2, g_3 violate transitivity so that

$$g_1 \succ g_2 \succ g_3 \succ g_1.$$

We can now give this person g_2 for free and then offer her to exchange g_2 plus a small amount x_1 of money for g_1. Given that she prefers g_1 to g_2 and that x_1 is, indeed, small, we can expect her to accept our offer. Next, we offer her to trade g_1 plus a small amount, say x_2, for g_3 which she prefers to g_1. Under similar assumptions as previously, it is plausible to expect her to accept. We then offer her to exchange g_2 for her g_3 plus x_3 units of money. Again we can assume that she accepts our offer. Now, the person is again in possession of g_2 which she had in the beginning, but she has lost $x_1 + x_2 + x_3$ units in exchanges. Moreover, we can repeat the cycle of pairwise acceptable offers and pump the person for her assets, certainly a pretty strong argument for trying to make sure that one's preferences are transitive.

The money pump is a dynamic construct. In other words, it involves successive trades. Another, more subtle, dynamic argument for the principles of EU theory is presented by Hammond (1988). Hammond argues that honouring certain plausible principles of choice in dynamic settings is *ipso facto* a commitment to the EU theory. Let us consider an example discussed by McClennen (1990). Figure 2.8 depicts a decision tree where nodes D^1 and D^2 are decision points, while nodes C^1, C^2, and C^3 are chance points, that is the continuation depends on some random event with known outcome probabilities. Thus, at the outset (in node D^1) the decision maker chooses either the upper or lower branch. If she chooses the latter, she gets a risky prospect where the outcomes are either 1 or 2401 with probabilities 66/100 and 34/100, respectively. If she chooses the upper branch, then with probability 66/100 she gets the payoff 0. With probability 34/100 she continues to the second decision node D^2 where she chooses between the certain prospect 2400 and the risky prospect $(0, 1/34; 2500, 33/34)$.

Consider now five risky prospects:

$r_1 = (2400, 1)$
$r_2 = (2500, 33/34; 0, 1/34)$
$r_3 = (2400, 34/100; 0, 66/100)$
$r_4 = (2500, 33/100; 0, 67/100)$
$r_5 = (2401, 34/100; 1, 66/100).$

Suppose that the decision maker has preferences that coincide with those of the majority of subjects in the certainty effect experiments of

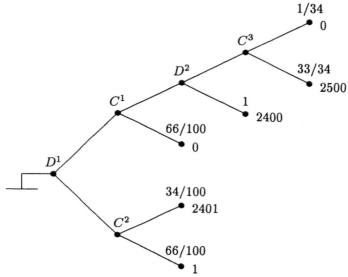

Figure 2.8: Decision Tree Example

Kahneman and Tversky, to wit, $r_1 \succ r_2$ and $r_4 \succ r_3$. It is natural to assume that $r_5 \succ r_3$ and that — in view of the extremely small difference between r_3 and r_5 — also $r_4 \succ r_5$.

Let us suppose that the decision maker chooses the upper branch at D^1 anticipating that, should she eventually reach D^2, she would choose the upper branch again thereby ending up with C^3, that is, r_2. This plan is plausible since the decision maker prefers r_4 to r_5. She thus rejects C^2, that is, r_5. On closer inspection this plan amounts to choosing r_4 instead of r_5 since a person choosing the upper branch first and then again at D^2 faces the risky prospect of getting 0 with probability $66/100 + 34/100 \cdot 1/34 = 67/100$ and 2500 with probability $34/100 \cdot 33/34 = 33/100$. This is precisely r_4.

Suppose now that the person sets out to follow her plan and chooses the upper branch at D^1. Suppose, moreover, that at chance node C^1 the lower branch will be chosen. This event has the probability of $66/100$. The payoff for the decision maker is then 0. Had she chosen the lower branch at D^1, the outcome that would have occurred with the same probability $66/100$ would have been 1, that is, an outcome preferred to 0. Suppose, on the other hand, that the person reaches the second decision node. We have assumed that the person prefers r_1 to r_2 and thus chooses the former alternative, ending up with 2400 with certainty. However, the same chance condition — an event with identical probability $34/100$ — would have given her 2401 had she chosen the lower branch at D^1. Thus, in both chance conditions — one occurring with probability 66/100 and the other with probability 34/100 — choosing r_5 gives better results than choosing the upper branch at D^1. Following the plan, then, seems to lead the decision maker with preferences that violate independence conditions to the choice of a dominated alternative.

Further reading: Elster (1979); Gauthier (1996; 1997); McClennen (1990).

2.12   DECISION THEORY AND INSTITUTIONAL DESIGN

Decision theory focuses on principles that guide individual decision making. Of particular interest are principles that characterize rational behaviour since people are often thought to be rational in the pursuit of their goals. Thus, in the design of institutions one should be able to predict what rationality of behaviour entails under various institutional arrangements. It is, however, known that not all human behaviour is, intuitively speaking, rational. Thus, one is led to ask whether the rationality of behaviour is but a guess in the design of institutions.

Some principles of rationality, for example the choice of a dominant alternative, are intuitively so compelling that any decision maker would presumably choose a dominant alternative if she knows that one exists. Moreover, one would expect any decision maker to change her mind if she is told that what she chose in a previous situation was a dominated alternative. There may, however, be circumstances in which other considerations than individual rationality suggest the choice of a dominated alternative. These are typically related to strategic contexts, that is, situations in which several decision makers make interdependent decisions while trying to anticipate each other's choices and knowing that the others are facing an analogous situation. We shall deal with those situations in the next chapter.

The dominance criterion is, however, a relatively straightforward principle of rationality. Its very straightforwardness makes it particularly suitable for the design of institutions. Starting from those goal states one wishes to achieve by constructing an institution, one is well advised to come up with such arrangements in which the intended states are results of dominant choices by those acting within the institution. Thereby one would guarantee that nobody would benefit from not acting according to the predicted principle. Also the EU criterion can be justified in similar terms. In other words, by resorting to alternatives satisfying the principle one can rest assured that in the long run one's payoff will not be smaller than that ensuing from any other course of action. It is thus plausible to predict that the actors learn to act according to the principles of rationality. If they act otherwise, they will lose in terms of long-run payoffs. In the case of the dominance criterion they will lose in short-term payoffs as well.

The study of individual decision making is thus important for the design of institutions since it enables us to predict what rational behaviour entails in various settings and, consequently, helps in assessing what each individual is likely to do in a given type of situation. However, decision theory focuses on an essentially passive environment. Since, on the other hand, economic, political and social institutions, by definition, involve a multitude of individuals, the question arises as to how the strategic aspects − anticipations and expectations regarding other people's behaviour − affect individual choice behaviour. These aspects are the domain of game theory, to which we turn after a few bibliographical comments.

2.13   BIBLIOGRAPHICAL REMARKS

The literature on decision theory is vast. For readable introductions the reader is referred to the works of Chernoff and Moses (1957), Luce and Raiffa (1957) and Fishburn (1970). Despite the fact that these books are old, they are still good guides for modern decision theory. The classic work in this field is von Neumann and Morgenstern's *magnum opus* (1944). French (1986) gives a more up to date general introduction to decision theory. Harsanyi's text has been extensively used in the preceding (1977).

The foundational problems of rationality are discussed for example by Binmore (1987; 1988) and Simon (1957; 1972; 1977). These authors discuss *inter alia* the nature of rationality (is a rational person maximizing something?), the 'givens' of individual choice theory (how is the alternative set formed?) and the dependence of various rationality paradoxes on the assumption that the individuals have unlimited information- processing capabilities.

Perhaps the most cited work in the theory of measurement is the three-volume treatise *Foundations of Measurement* (Krantz et al. 1971; 1989; 1990). A considerably more accessible, albeit less comprehensive, account is the book by Roberts (1979).

Allais's paradox and other empirical deviations from the EU theory have provided the motivation for several international conferences, for example the biannual conference Foundations and Applications of Utility, Risk and Decision Theories (FUR). Its proceedings contain much of the recent empirical and theoretical work on prospect theory, preference reversal and regret theory (Allais and Hagen 1979; Munier 1988; Munier and Shakun 1988). Also the interesting exchange between Hammond and McClennen on dynamic consistency took place under the auspices of FUR. Another regular conference, namely the conference of Subjective Probability, Utility and Decision Making, ought to be mentioned. Its proceedings have been published in alternating fashion as separate conference proceedings volumes and as special issues of *Acta Psychologica*. Also the journals *Theory and Decision* as well as *Journal of Risk and Uncertainty* regularly publish articles on the foundations of decision theory.

Alternatives to EU theory have been proposed, notably by Allais (1979), Fishburn (1988) and Machina (1982). Possible explanations and implications of dropping the assumption that the decision makers have transitive preference relations have also been explored in the literature (Aizerman and Aleskerov 1995; Cowan and Fishburn 1988; Bar-Hillel and Margalit 1988; Nurmi 1991). Probabilistic choice theories replace the conditions on preference relations by various consistency requirements concerning the choices to be made from the subsets of alternatives (Coughlin 1992; Intriligator 1982; Luce 1959).

Experimental work on EU theory violations involving political decision alternatives are reported by Quattrone and Tversky (1988) and Herne (1997a; 1997b).

3 Games

The setting of game theory is the same as that of decision theory (see Figure 2.1 on page 6) with the important difference that in the former the environment consists of other actors or players, with their own interests concerning the outcomes and — *eo ipso* — consequences. The development of game theory has gone through phases of compartmentalization (specialization) and integration. The intention of the present chapter is not to capture all or even the most important results achieved in the theory of games during its half-century-long history. Rather the aim is to outline the conceptual apparatus of the theory, its peculiar way of modelling human interaction and its role in institutional design. Thus the approach of the present chapter is highly selective. For more general and relatively up-to-date treatments the reader is referred to Fudenberg and Tirole's as well as Morrow's works (Fudenberg and Tirole 1991; Morrow 1994).

Many important aspects of political institutions (party systems, voting procedures, legislative systems) can be modelled as games involving actors, strategies and outcomes. The contribution of game theory to institutional design is to outline the likely outcomes in various kinds of games. These predictions help us in assessing the merits of the institutional arrangements underlying those games.

The basic aim of institutional design is to find arrangements in which rational behaviour results in outcomes that are desirable. The game-theoretic predictions are typically called solutions. A common feature of them is that the solution outcomes are equilibria, that is, stable in various meanings of the term. Stability, in turn, is understood as a state that is immune to change, again in a specific sense. Game theory gives us various equilibrium notions, motivates them in terms of player rationality and sometimes even suggests methods of reaching those outcomes. To the extent that we are able to characterize the institutions in game-theoretic terms, we can thus take advantage of game theory in predicting what will be the likely outcomes in the institutions we are focusing upon.

The system of game theory can be classified in many ways (Intriligator 1971; Riker and Ordeshook 1973):

- according to the number of players. The main classes are:
  1. two-person games
  2. n-person games

- according to the nature of the payoff function, that is, the function that maps the k-tuples of individual choices into payoffs to each of the k players. The main classes are:
  1. constant-sum games
  2. non-constant-sum games

- according to the number of choices available to each player. The main classes are:
  1. finite games
  2. infinite games

- according to the possibility of making binding commitments before the game is actually played:
  1. cooperative games
  2. non-cooperative games

A few comments on these classes are worth making. The first classification coincides with a difference in modelling apparatus. In two-person games, one typically looks at the various choice combinations of the two players and tries to determine the likely outcomes ensuing from the game, while in n-person games the focus is on player coalitions and the division of payoffs within them. The difference has implications for modelling devices as well: in two-person games one often resorts to matrix representation in which one player's choices − say l in number − are modelled as rows and the other player's choices − say m in number − as columns of an $l \times m$ matrix. This is the matrix- or strategic-form representation of the game. Alternatively, a two-person game may be represented as a game tree or in extensive form. In this representation the game is described by depicting all possible sequences of moves by each player. Although both matrix and extensive forms could in principle be used in describing games with more than two players, their primary field of application is for practical reasons in two-person games. In n-person games, on the other hand, one typically resorts to characteristic function representation. This involves indicating for each 2^n coalitions that can be formed of n players, the payoff that this coalition gets should it be formed in the course of play. The first classification thus relates to the modelling apparatus that is used in defining and analysing games.

A terminological remark should be made in this context, namely sometimes a distinction has to be made between the concepts of choice and strategy. By the former one means the basic constituent of a strategy, that is, a strategy consists of choices arranged in a sequence. The latter refers to a rule that specifies which choice is to be made in all conceivable contingencies of the game. Thus a strategy gives an unambiguous instruction about the choice in all possible stages of the game. Often the distinction between strategy and choice is of no consequence and the two concepts can then be used interchangeably.

The second classification has to do with the degree to which the players' interests coincide. In constant-sum games — a special case being zero-sum games — the players have to divide a constant amount of payoff among themselves. In two-person constant-sum games this inevitably means that the more one player gets, the less is left to the other. This gives these games a particularly antagonistic outlook. In n-person constant-sum games the conflict between players may become more blurred since it is now a larger set of players whose payoff sum has to be a fixed constant in all outcomes. However, also in non-constant-sum games where by definition the sum of payoffs of all players is not always the same constant, there may be considerable conflict between players. The non-constant-sum nature only guarantees that the 'cake' to be divided among players is not always of strictly equal size.

Whether the game is finite or infinite is of limited significance to our discussion. We shall be dealing primarily with finite games. The infinite games have applications in situations where the n players make their choices from a set of continuous functions, for example time functions, and where the outcomes are continuous flows of payoff determined by the value n-tuples of those functions. In the games we shall be discussing the players have a finite and usually rather small set of choices at their disposal.

The cooperativeness of a game is determined by the possibility of players to make binding commitments about their choices before the actual play of the game. Those games in which such a commitment is possible are cooperative, others are non-cooperative. It is clear that non-cooperative games are more general than cooperative ones since making a commitment can be viewed as a strategy in a non-cooperative game. The reduction of a cooperative game into a non-cooperative one may, however, involve the enlargement of the player set to include the possible contract-enforcing parties with their interests.

Our coverage of these games is highly selective. We shall begin with the theoretically best-known class of games, namely two-person zero-sum ones.

C UN	not produce (NP)	produce (P)
not inspect (NI)	0	−1
inspect (I)	−1	+1

Table 3.1: The Inspection Game

3.1 CONSTANT-SUM GAMES

As an example of two-person zero-sum game consider a situation where the United Nations is conducting inspections in a country C suspected to be producing weapons banned by international treaties. Let us assume that the leaders of C have an interest in building such weapons for eventual use against their neighbouring countries. The leaders of C would like to produce the weapons undetected, while the UN would like to stop the production. A necessary condition for stopping is the detection of production. The UN personnel have made up a list of industrial facilities where the production of banned weapons could conceivably take place. The inspection is costly and, therefore, the UN personnel would prefer not to inspect sites where no production takes place. Consider now a fixed industrial facility F in the list of suspicious sites that has not yet been inspected. The UN personnel have basically two options: either to inspect now or postpone the inspection indefinitely. The leaders of C, on the other hand, have two choices regarding F: either to use it now for the production of banned weapons or to use it for some legitimate purpose. The situation can be described by means of Table 3.1.

The table indicates the payoffs for the UN under various choice combinations of players. If the UN inspects and finds that production of banned material is under way in the facility, its payoff is +1, while the payoff of C is −1. In zero-sum games it is common not to mention explicitly the payoffs of the player whose choices are represented by column (the column player or Column, for brevity) since his payoffs can be inferred from the payoff of the other player (the row player or Row). This convention is followed here as well. If the UN does not inspect and the facility is being used for production of banned material, then the payoff for the UN is −1. The same payoff accrues to UN if it inspects a facility that is not used for production. If no inspection is made in a facility that is not used for production, then the payoff to the UN is 0. It is assumed that both players know their own payoffs and those of the other player. They are also assumed to know that the other player knows that they know and so on *ad infinitum*.

What is the likely outcome of the game? By invoking the dominance criterion of rationality we can build our predictions on the following possibilities:

- Both players have a dominant strategy, that is, a choice that dominates all other choices in the sense of bringing at least as high a payoff than any other choice in all contingencies and, for any other choice, bringing a strictly higher payoff than the latter in at least one contingency. We can predict that, should the players have a dominant strategy, they will choose it since nothing can be gained by not choosing it.
- One of the players has a dominant strategy. Then we can predict on the same grounds as above that he will choose it. Since we also assume that both players know the payoff matrix, we can predict that the player who does not have a dominant strategy knows that the other player has one. Consequently, the player without a dominant strategy can predict that the other chooses his dominant strategy, leaving him (the player without a dominant strategy) to choose from among those choices that are best replies to the dominant strategy. Thus we have a prediction which yields a unique outcome provided that each entry in the matrix is different from the others (no ties).
- There is a Nash equilibrium outcome, that is, an outcome such that, should it be reached, neither player would benefit from unilateral departure from it. More precisely, a Nash equilibrium in pure strategies is an outcome resulting from the row player's choice of a_i and the column player's choice of b_j iff a_i is the best reply to b_j and *vice versa*. In other words, in a Nash equilibrium neither player regrets his choice, given the other player's choice. If both players have a dominant choice, then the outcome is obviously a Nash equilibrium. Also if one player has a dominant strategy and the other chooses the best reply to it, the outcome is a Nash equilibrium. However, an outcome may be a Nash equilibrium without there being a dominant choice for either player.

Although the dominance criterion of rationality − whenever you have a dominant alternative, choose it − is the strongest conceivable norm of choice, there are situations in which one may hesitate to act according to it. We shall encounter such situations later on, but in the present example there would be nothing to prevent a player from making a dominant choice if there were one. But neither player has one. Thus the first two possibilities above do not help us in predicting the likely outcome. Nor does the third, since all outcomes are vulnerable to unilateral second thoughts. The outcome +1 (yielding −1 to C) would be vulnerable to C's regret since the outcome −1 (giving +1 to C and ensuing from his unilateral choice of NP given the choice of I by the UN) would clearly be better. In the outcome -1 in the lower left-hand corner of the matrix the UN, in turn, would regret and would rather choose NI, given C's choice of NP. Similarly, in outcome -1 in the upper right-hand corner of the matrix the UN would rather choose I if it knew that C chooses P. Finally, in outcome 0, C would rather choose P than NP. Thus none of the outcomes is a Nash equilibrium.

One of the most important results in the theory of games states that in all finite strategic-form games there is at least one Nash equilibrium in either pure strategies or mixed strategies (Nash 1950). Mixed strategies are probability mixtures of pure strategies. Thus, for example, $(P, p; NP, (1 - p))$ could be viewed as mixed strategy calling for the choice of P with probability p and the choice NP with probability $1 - p$. The result is very general. It covers not only the two-person zero-sum games, but all finite games in strategic form, that is, in the form where each player is assumed to make a choice from a finite set of alternatives and where the choices jointly determine the outcomes.

Since the Inspection Game does not have a Nash equilibrium in pure strategies, it must have at least one such equilibrium in mixed ones. In other words, there must be a pair of mixed strategies, one for the UN and one for the leaders of C, so that if one of them chooses his element of the pair, the other cannot benefit from choosing any other strategy than his element of the pair. It should be observed that when dealing with mixed strategies the payoffs are expected values. Thus, in a mixed strategy Nash equilibrium neither player benefits in terms of expected values of payoffs from choosing otherwise.

To find out the mixed-strategy Nash equilibria in the Inspection Game, let us first look at the admissible transformations of the payoff matrix. These are defined as those permutations (switchings) of rows and columns that leave the strategic properties of the game unaffected. The strategic properties of the game remain unaffected if the players are faced with identical uncertainty (strategic uncertainty) regarding the outcomes as before. The strategic uncertainty pertains to the outcomes that may ensue when a choice is made. Thus, for example, in the Inspection Game the strategic uncertainty of the UN is that when it chooses I, it may end up with either $+1$ or -1, and when it chooses NI, its payoff may be either -1 or 0. This uncertainty is unaffected if row I with its entries is placed above row NI. Similarly, C's strategic uncertainty remains the same if the column NP is placed to the right side of column P.

Two admissible transformations of the Inspection Game are presented in Table 3.2. The last version of the game is one in which the

C		
UN	P	NP
NI	−1	0
I	+1	−1

$\rightarrow$

C		
UN	P	NP
I	+1	−1
NI	−1	0

Table 3.2: Admissible Transformations of the Inspection Game

highest payoff of Row is in the upper left-hand corner and the next highest in the lower right-hand corner of the matrix. Not all 2 × 2 games — that is, games in which each player has two choices — can be rendered into such a configuration. A moment's reflection reveals that in those games where this configuration cannot be reached, we have a Nash equilibrium in pure strategies. To wit, it is always possible to transform the matrix so that the highest payoff of Row is in the upper left-hand corner. If the next highest is not in the lower right-hand one, then it has to be either in the second row and first column or in the first row and second column. In the former case, Column has a dominant strategy, namely the right column, since Column's highest and second highest payoffs must be located in that column. In the latter case, Row has a dominant strategy, namely the upper row. In both cases, then, one player has a dominant strategy and the best response to it by the other player results in a pure-strategy Nash equilibrium.

However, in the Inspection Game there are no dominant strategies and we are able to reach a configuration of the right-most matrix in Table 3.2. For any probability mixture of I and NI by Row, we are able to compute the expected value of the payoff if Column always resorts to P. Similarly, we are able to compute the expected value of the payoff for Row if Column uniformly uses NP. By the zero sum property the corresponding values for Column can be directly inferred. For example, if Row chooses I with probability 1/4 and NI with probability 3/4, then Row's expected payoff is $1/4 \cdot 1 + 3/4 \cdot (-1) = -1/2$ if Column chooses P and $1/4 \cdot (-1) + 3/4 \cdot 0 = -1/4$ if Column chooses NP. The corresponding expected values for Column are, of course, 1/2 and 1/4.

Thus, if Column knew the mixed strategy of Row, he would have an incentive to choose P rather than NP. But surely Column can randomize his choices as well. What Row can be assured of is that by using the probability mixture $p_{1/4} = (I, 1/4; NP, 3/4)$ his expected payoff is no less than $-1/2$. This is what Row can unilaterally guarantee himself in expected payoffs when he resorts to the mixed choice $p_{1/4}$. This can be viewed as the security level of strategy $p_{1/4}$. It is noteworthy that the randomized choice $p_{1/4}$ is accompanied by a higher security level than either of the pure choices I and NI.

The value of the game is the security level payoff related to the maximin strategy, that is, it is the minimum payoff that can be obtained by making the choice which has the largest security level. The value of the Inspection Game for Row is -1 if only pure choices can be made. However, as we have just seen, this value for Row can be increased when randomized choices can be used. It is also worth noticing that Row's mixed strategy brings joint benefits for both players of this zero-sum game. To wit, if Row resorts to the mixed strategy $p_{1/4}$ Column's value increases from the security level 0 to $1/4$.

The randomized choice $p_{1/4}$ is, however, not the best Row can do in terms of expected payoffs. Consider $p_{1/3} = (I, 1/3; NI, 2/3)$. This gives Row the expected payoff $-1/3$ both when Column chooses P and when Column chooses NP. In other words, by resorting to $p_{1/3}$ Row is able to guarantee himself a payoff $-1/3$ regardless of Column's choice since obviously a weighted average of two identical numbers is that very number. The payoff $-1/3$ is the value of the game for Row, that is, the highest payoff that he can guarantee himself unilaterally. That this value is, indeed, the maximum can be see by considering

- randomized choices giving I somewhat higher probability than $1/3$, say $1/3 + \epsilon$ and, consequently, NI somewhat smaller probability $2/3 - \epsilon$, and
- randomized choices giving I somewhat lower probability than $1/3$, say $1/3 - \epsilon$ and, consequently, NI somewhat higher probability $2/3 + \epsilon$.

Consider the first case. If Column resorts to NP, the expected value for Row is $-1/3 - \epsilon$, that is, smaller than $-1/3$. In the second case if Column chooses P, the expected value for Row is $-1/3 - 2\epsilon$, thus demonstrating that $p_{1/3}$ yields the maximum value for Row.

From Column's viewpoint the computations are analogous. In other words, one considers probability mixtures $q = (P, q; NP, 1 - q)$ and determines the expected payoffs. The value of the game is determined by solving for q in the equation:

$$1 - 2q = q.$$

The randomized choice $q_{1/3} = (P, 1/3; NP, 2/3)$ makes Column's expected payoff independent of Row's choice. It also yields the maximum payoff for Column, that is, $1/3$ which is thus the value of the Inspection Game for the column player.

The mixed strategies $p_{1/3}$ of Row and $q_{1/3}$ of Column thus maximize the security level payoffs for both players. By construction, these payoffs are independent of the other player's choice. Hence, if Row knew that Column was going to resort to $q_{1/3}$ he could not benefit from making any other choice but $p_{1/3}$ since $q_{1/3}$ gives Column 1/3 no matter what Row chooses. *Ipso facto* $q_{1/3}$ gives Row $-1/3$. Similarly, Column could not benefit from choosing anything else but $q_{1/3}$ if he knew that Row chooses $p_{1/3}$. Thus these two randomized choices constitute, indeed, a Nash equilibrium.

Some games have Nash equilibria in pure strategies, some in mixed ones and some in both. The concept of mixed strategy, however, calls for a few remarks. Firstly, under the classic frequentist intepretation of probability a mixed strategy makes sense only in long sequences of repetitions of the game. In those contexts one may interpret $p_{1/3}$ as an instruction to use a random device with outcome probabilities 1/3 and 2/3 — for example outcomes five or six on the one hand, and all other outcomes, on the other, in a throw of an unbiased dice — and make the choice of I or NI depending on the outcomes. Thereby one could be assured that in a long sequence of similar decision situations one's expected payoff would coincide with the value of the game for Row in the Inspection Game. But suppose that the UN authorities are looking at a particular facility in the list of suspicious sites. The question then arises as to what meaning the mixed choices have in this particular case. Very little, since, regardless of the random mechanism underlying it, the choice is either I or NI by the UN. Hence inevitably the payoff for the UN is one of the entries in Table 3.1 and its opposite number is the payoff of C. However, in a long sequence of similar inspection situations, the equilibrium strategy for the UN is to inspect 1/3 of the sites and for the leaders of C to produce banned material in 1/3 of the facilities.

Further reading: Brams (1975); Hamburger (1979); Morrow (1994).

3.2 NASH EQUILIBRIA IN 2×2 GAMES

A mixed-strategy Nash equilibrium outcome in a two-person game consists of two probability vectors, one for each player. In 2×2 games, that is, games in which both players have two pure choices, a mixed strategy Nash equilibrium can be represented as a pair $(p*, q*)$ of probabilities, $p*$ indicating that Row's probability of choosing the higher row is $p*$ and his probability of choosing the lower row $1 - p*$. Similarly, $q*$ is Column's probability of choosing the left column and $1 - q*$ his probability of choosing the right column. What makes this pair of probabilities a Nash equilibrium is that once one of the players has chosen his mixed strategy, the other cannot benefit from choosing other than his mixed strategy.

In two-person 2×2 zero-sum games the method of finding the mixed

Column Row	left	right
up	R_{11}, C_{11}	R_{12}, C_{12}
down	R_{21}, C_{21}	R_{22}, C_{22}

Table 3.3: A Non-Constant Sum Game

strategy Nash equilibria outlined in the preceding section takes its point of departure in making one player's expected payoff independent of the choice of the other player. In fact, the probability mixture of choices is determined on the basis of this condition. Thus, for example, Row's equilibrium strategy probabilities $p*$ and $1 - p*$ for choosing the upper and lower row, respectively, are determined by the condition that regardless of what Column chooses, Row's expected payoff will be the same. But in zero-sum games this necessarily also means that Column's expected payoffs are the same regardless of which column he chooses.

Consider Table 3.3. This game is not necessarily zero-sum, but each cell indicates the payoff to Row and Column. Thus, for example, if Row chooses 'down' and Column 'left', the payoffs are R_{21} to Row and C_{21} to Column. In a zero-sum game, the condition for p is that

$$pR_{11} + (1 - p)R_{21} = pR_{12} + (1 - p)R_{22}.$$

Now when this is the case, then we also know that

$$pC_{11} + (1 - p)C_{21} = pC_{12} + (1 - p)C_{22},$$

since the game is zero-sum.

When the game is not zero-sum, the value of p that is a solution to the former equation is not in general the value that solves the latter equation. Moreover, if we solve for p in the first equation, we typically get a value that is not an equilibrium, since the value of p that makes Row indifferent as to whether Column chooses 'left' or 'right' may well be such that Column has a clear preference between the columns. Consequently, if Column knows that Row chooses a mixed strategy as a solution to the first equation, his best response to that is not a mixed strategy that makes Column indifferent between Row's choices. His best response is the one that maximizes his expected payoff, given Row's mixed strategy. Thus the situation looks very much like the one we started from: the best response depends on the choice of the opponent.

Yet Nash's result says that there is at least one equilibrium in all finite games. How can we find it in the more general 2×2 games? The answer is found by looking again at the definition of the equilibrium and paying attention to the fact that the condition for equilibrium values of p and q is that, say, Row cannot take advantage of the knowledge that Column is using his equilibrium strategy. Thus Column's strategy mixture is such that Row is indifferent between 'up' and 'down'. In Table 3.3 the equilibrium probabilities are solutions to the following two equations:

$$pC_{11} + (1 - p)C_{21} = pC_{12} + (1 - p)C_{22}$$

$$qR_{11} + (1 - q)R_{12} = qR_{21} + (1 - q)R_{22}.$$

The solution values are:

$$p = \frac{C_{22} - C_{21}}{C_{11} + C_{22} - C_{21} - C_{12}}$$

$$q = \frac{R_{22} - R_{12}}{R_{11} + R_{22} - R_{12} - R_{21}}.$$

In a mixed-strategy Nash equilibrium, then, the probability that Row chooses the upper row is determined by Column's payoffs and the probability that Column chooses the right column is determined by Row's payoffs. This is certainly not surprising, given the construction underlying the computation of the probabilities. It has, however, rather peculiar implications in situations where one resorts to the Nash equilibrium in institutional design. To wit, if one aims at an institution in which the desired forms of behaviour result in Nash equilibria, then the only way to change Row's behaviour is to change Column's payoffs. This observation is conditional on there being just mixed-strategy Nash equilibria. But under this condition, the implication contradicts the conventional wisdom according to which to encourage a person to act in certain ways one should − through rewards and/or punishments − modify his payoffs. Building his case largely on the computation of mixed-strategy Nash equilibria George Tsebelis (1989) has criticized the use of this conventional wisdom in game environments. We shall return to his argument later on.

Further reading: Brams (1975); Holler (1993); Tsebelis (1989).

3.3 DOMINANCE SOLVABILITY

Nash's result tells us that equilibria exist in all finite games and that, once reached, they are characterized by a certain stability, to wit, no unilateral movement away from the equilibrium outcome benefits any player. In two-person games where at least one player has a dominant strategy and the other player chooses the best reply to this strategy, the outcome is a Nash equilibrium. But as we have seen, Nash equilibria also exist in games where neither player has a dominant strategy. In those games the predictive value of the Nash equilibrium is more limited since even though the outcomes, once reached, may be stable, we have no assurance that the players 'find' the outcome. Another problem pertaining to the Nash equilibrium is the fact that there are games with several equilibria. Indeed there may be infinitely many of them, as in the following game introduced by Ariel Rubinstein (1982).

Example. We consider a two-person bargaining game where a 'pie' is to be partitioned among players. If the players can agree upon a division, it is implemented. Otherwise, they get nothing. We assume that each player prefers a larger piece to a smaller one. The set S consists of all real numbers between 0 and 1. Then the possible solutions of the game can be represented as values $s \in S$ so that s is the portion that player 1 gets and $1 - s$ the portion player 2 gets. In this game any of the infinitely many $(s, 1 - s)$ pairs is a Nash equilibrium since clearly once player 1 has proposed a solution giving him s, the best player 2 can do is to accept $1 - s$. Similarly, if player 2 has the chance to propose a solution, say a for himself and $1 - a$ for player 1, then the best reply of player 1 is to accept it. Thus there are infinitely many Nash equilibria in this game.

A finite-game example of the multiplicity of Nash equilibria is given in Table 3.4 which shows that in some 2×2 games all outcomes may be Nash equilibria in pure strategies. This is not very helpful if one aims at designing or evaluating institutions on the basis of the behaviour of individuals in the equilibrium. Consequently, many refinements of Nash equilibria have been suggested in the literature. These are solution concepts that select a subset of Nash equilibria. One particularly plausible solution concept of this nature is based on dominance solvability of the game. To introduce this idea we define the concept of strong dominance.

Definition 3.1 *A strategy s_i of player I strongly dominates another strategy s_j of the same player, iff s_i leads to a strictly better payoff for I than s_j for each strategy choice of the other players.*

In other words, given any combination of strategies of other players a strongly dominating strategy is not only as good as, but strictly better than, the strategy it dominates. The idea underlying dominance solvability is that eliminating strongly dominated strategies — that is, strategies that are dominated by one or more other strategies — is a rational thing to do. Thus, in looking for outcomes that are likely to result from rational play, one is well-advised to omit dominated strategies. Intuitively

Column Row	left	right
up	2, 3	3, 3
down	2, 1	3, 1

Table 3.4: Too Many Nash Equilibria

it would seem that eliminating weakly dominated strategies would be an almost equally rational thing to do. It certainly helps in simplifying games, but it is a matter of some controversy whether one should eliminate weakly dominated strategies in finding solutions to games. Ken Binmore's example (Table 3.5) illustrates the procedure (Binmore 1992).

In some games the process of eliminating dominated strategies can be repeated so that after a few stages one ends up with one strategy for each player. When this is the case, the game is called dominance-solvable. After all, what one is left with is a single outcome, that is, a solution. In Table 3.5 there are three Nash equilibria in pure strategies, one resulting from Row's 'up' and Column's 'right', one resulting from Row's 'down' and Column's 'left' and one resulting from Row's 'down' and Column's 'far right'. Upon inspecting the players' strategies one observes that neither of Row's strategies strongly dominates the other, but Column's 'far right' strongly dominates his 'far left'. Consequently, we may eliminate the 'far left' column, whereupon we notice that Column's 'far right' weakly dominates both 'left' and 'right'. Consequently, after also eliminating the weakly dominated strategies of Column, we end up with a matrix with just one column, namely 'far right'. In this matrix, of course, Row's 'down' trivially dominates 'up' and the solution of this game is 'down' and 'far right' outcome $(3, 2)$.

Had we eliminated only strongly dominated strategies, our prediction would have been ambiguous. In fact, only one column would have been eliminated. By also casting out weakly dominated strategies we were eventually left with just one of the original three Nash equilibria. It is no accident that the outcome surviving the elimination process is a Nash equilibrium. However, as we have just seen, not all Nash equilibria survive the process.

There is another notion of strong dominance in the literature which is worth mentioning since it takes − albeit indirectly − the mixed strategies into account in eliminating dominated strategies. It considers a whole classification of the opponent's mixed strategies, for example strategies according to which the opponent makes a given choice with at least a $1/3$ probability and those according to which he makes this choice with

Column Row	far left	left	right	far right
up	1, 0	1, 0	2, 3	2, 3
down	0, 1	3, 2	0, 1	3, 2

Table 3.5: A Dominance-Solvable Game

at most a 1/3 probability, and shows that a given strategy of the player never maximizes the payoff.

Definition 3.2 *Player I's strategy s_i is strongly dominated if I has another strategy s_j that gives I a strictly higher payoff for any combination of choices of other players.*

To see how this concept of strong dominance — and the dominance solvability concept that can be derived from it — differs from the concept discussed in the preceding, consider Adam Brandenburger's example in Table 3.6 (Brandenburger 1992).

In the sense of the first definition, none of the strategies strongly dominates any other strategy. In fact, all strategies are weakly undominated. Thus this game is not dominance-solvable. Consider now the second notion of strong dominance. While both Row's strategies are still strongly undominated, Column's strategy 'right' gives Column a strictly smaller payoff than 'left' against all Row's strategies that assign 'up' a (subjective) probability higher than or equal to 1/2. Similarly, Column's strategy 'right' gives a payoff that is strictly less than 'centre' would give him against all Row's strategies in which 'up' is chosen with probability less than 1/2. Thus, in all possible cases some other strategy than 'right' maximizes Column's payoff. Consequently, according to the latter definition 'right' is strongly dominated. Once 'right' is eliminated, we observe that Row's 'down' is dominated by 'up', whereupon 'up' — 'left' emerges as the solution outcome, giving Row 1 and Column 10.

In institutional design dominance solvability has a particularly prominent place since it, along with the notion of backwards induction, has found interesting applications in the analysis and evaluation of voting institutions. In contrast to dominance solvability, the backwards induction principle can be explained only in the context of extensive-form games.

Further reading: Binmore (1992); Brandenburger (1992); Fudenberg and Tirole (1991).

Column Row	left	centre	right
up	1,10	3,0	1,3
down	0,0	2,10	10,3

Table 3.6: Another Concept of Strong Dominance

3.4 ZERMELO'S ALGORITHM

Extensive-form games describe games as trees in which each player's choices are represented by branches originating from nodes which have been reached as a result of previous choices of players who have made their choices earlier in the game. Extensive-form representations include all possible choice sequences of players, that is, all possible game histories. Although in many contexts the matrix- and extensive-form representations are interchangeable and all extensive-form two-person games can be represented in a matrix form, there are contexts in which the extensive form is essential for understanding what rational play would imply.

In the late 1970s Henry Hamburger discussed a game that about ten years later turned out to be a strikingly accurate description of a real world conflict, namely the Gulf War (Hamburger 1979). In that game a Sabre Rattler country plans to invade an Oil Producer country. The decision to invade is denoted by branch a emanating from the SR (Sabre Rattler) node. The decision to maintain the status quo is denoted by the branch labeled b. In the deliberations of Oil Producer the possibility of blowing up the oilfields is one option (branch a), provided that the invasion takes place. Not to blow up the fields is denoted by branch b emanating from the OP (Oil Producer) node. The entire extensive form of the game is depicted in Figure 3.1. The payoffs resulting from each sequence of choices are indicated in parentheses so that Sabre Rattler's payoff is before the comma.

In view of the nature of conflict we can make certain general assumptions about the order of magnitude of the payoffs for each player. Thus, in order for Oil Producer's threat to blow up the fields if attacked to have any effect at all on Sabre Rattler, we have to assume that $E > A$, that is, Sabre Rattler's payoff is larger in the status quo than if it attacks Oil Producer and the latter blows up the fields. On the other hand, $C > E$ also seems clear in view of the fact that Sabre Rattler ponders upon attacking Oil Producer in the first place. Similarly, B is presumably Oil Producer's worst payoff and, in particular, worse than D.

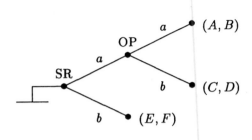

Figure 3.1: The Sabre Rattler Game

These assumptions about payoffs enable us to use backwards induction in predicting what rational players would end up with in this game. Starting from the final outcomes, we observe that if Oil Producer has any effect at all on the outcomes, that is, if his turn ever comes in this game, he determines whether the outcome is (A, B) or (C, D). Since he prefers D to B, we may conclude that if Oil Producer's turn comes, the outcome is (C, D). Thus, Sabre Rattler knows that if he chooses 'invade', the outcome is (C, D). On the other hand, if Sabre Rattler chooses 'not invade', the outcome is (E, F). Since he prefers C to E, he chooses 'invade'. The solution ensuing from backwards induction is, thus, that Sabre Rattler chooses 'invade' and Oil Producer thereafter 'not blow up', whereupon outcome (C, D) ensues.

The method used in the above analysis of the Sabre Rattler game is called backwards induction or Zermelo's algorithm. It was first outlined and discussed by Ernst Zermelo (1913). In operations research the method goes under the name of dynamic programming. Backwards induction proceeds from the end points or, more precisely, from the decision nodes immediately preceding the payoff points, of the extensive-form game towards the beginning. The analysis goes in the opposite direction to the 'arrow of time'. At each node on the way, one singles out the best outcome from the viewpoint of the player whose turn it is to make a choice at that node. For expositional purposes a new game tree may be constructed in which the decision nodes are replaced by outcomes singled out in the process of backwards induction. The node at which the game begins thus indicates the payoffs to players who resort to backwards induction. Figure 3.2 illustrates the application of Zermelo's algorithm to the Sabre Rattler game. Looking at the initial node, we immediately see the outcome of the game if the players resort to backwards induction.

Further reading: Binmore (1992); Brams (1975); Morrow (1994).

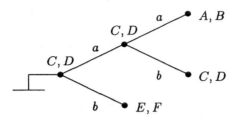

Figure 3.2: Applying Zermelo's Algorithm

3.5 SUBGAME-PERFECT EQUILIBRIA

Zermelo's algorithm provides a useful prediction of what rationality would entail in extensive-form games. In particular, it would be difficult to sustain outcomes that are not results of backwards induction. Consider a situation in which we are called upon to evaluate an institution that can be described using an extensive-form game. Let us assume that the outcomes generally considered fair or just are eliminated in the process of backwards induction. In this situation it is likely that the actors never reach the intended fair or just outcomes. Thus, to the extent that the institution is supposed to lead to those outcomes, it is likely to fail.

Similar argumentation can, of course, be presented regarding Nash equilibria. However, as was pointed out above, the Nash equilibria may simply be too numerous to be of predictive value. Moreover, there are examples of extensive-form games in which Nash equilibria call for downright implausible or irrational behaviour. Consider the two-person game presented in Figure 3.3 in extensive form and in Table 3.7 in matrix form (Baird et al. 1994, 64−5). It describes the setting of making a debt contract when contract enforcement is costly. More specifically, the game is initiated by Lender (L) who may or may not make a loan of 100 units to Debtor (D). If Lender does not make the loan (branch b), the game ends and the status quo prevails, that is, both players get 0 payoff. If the loan is made (branch a), Debtor may repay it (branch a) or default (branch b). In the former case, we assume that Debtor has earned 110 units, that is, in net terms 10 units, with the loan and that the interest Lender charges is 5 units. So, if Lender makes a loan and Debtor repays, both get 5 units. If Debtor defaults, Lender may (branch a) or may not (branch b) sue him. We assume that in the litigation process Lender would prevail, but the costs are extremely high. So in the case of suing, both end up 120 units poorer than in the beginning. If Lender does not

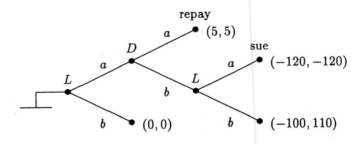

Figure 3.3: Debt Contract Game: Extensive Form

sue, he loses his 100 units loan and Debtor has all the benefit from its use.

In transforming this game into matrix form we start with the observation that Debtor has only two choices, default or repay. Lender, on the other hand, has four strategies since he has two choices in two information sets: (make loan, sue), (make loan, do not sue), (do not make loan, sue), (do not make loan, do not sue). We shall denote these strategies (MS), (MD), (DS) and (DD), respectively. The bold-face entries in the matrix form (Table 3.7) are Nash equilibria, that is, neither player can benefit from unilaterally deviating from them, given the other player's choice. The $(0, 0)$ outcomes result from Lender's not making the loan. One of these outcomes is the result of backwards induction. To wit, starting from the last decision node in the extensive form, we see that Lender has to choose between -120 and -100, the former payoff ensuing from his choosing to sue and the latter from his not suing. Clearly, -100 is better for Lender. Given this outcome, Debtor knows that, were he to default, Lender would not sue and thus Debtor would get 110, a far better payoff than 5, which he would get if he repaid the loan. Thus Lender knows that if he makes the loan, Debtor will default in the anticipation that Lender will not sue and Lender ends up with a loss of 100 units. Clearly, 0 is better and, hence, the result of backwards induction is $(0, 0)$, whereby the loan is not made.

But as Table 3.7 indicates, there is also $(5, 5)$ Nash equilibrium. It results from Lender choosing (make loan, sue) and Debtor choosing to repay. This equilibrium would, however, be implausible if the game ever came to the last decision node since there it would call for Lender to sue even though doing nothing would bring him a larger payoff. By the same token, the Nash equilibrium calling for Lender, first, not to make the loan and then, in the event of default, to sue, is implausible.

Let us make these intuitive remarks slightly more precise. An information set is a set of nodes that are indistinguishable to the player who is

Debtor Lender	repay	default
MS	**5,5**	−120,−120
MD	5,5	−100,110
DS	0,0	**0,0**
DD	0,0	**0,0**

Table 3.7: Debt Contract Game: Matrix Form

making the choice. For example, if a player has to choose without knowing the previous choice of the other player and the latter has two choices, there are two nodes in the first player's information set. At the time of choosing, he does not know in which one he is located. A subgame is a part of an extensive-form game that is obtained by considering the continuation of the game tree from a given node and ignoring all nodes and branches that precede it. All elements of any information set must, however, belong to the subgame, that is, if any node of an information set belongs to a subgame all other nodes of the same information set belong there as well. If the original game is excluded from the set of subgames, the remaining subgames are called proper ones. The proper subgames of the game of Figure 3.3 are the game starting from node 'D' onwards and the other game starting from the latter node labelled 'L' where the Lender chooses between suing and doing nothing. It is in this latter subgame that some Nash equilibria fail to suggest reasonable choice. In other words, some strategies that lead to Nash equilibria are not necessarily subgame-perfect.

Thus, not only are the Nash equilibria sometimes too plentiful, but they may even be implausible as guidelines for rational play. In the above example the particular problem with this equilibrium concept is that it dictates a choice in the game that does not make sense in various subgames that one might have to face for one reason or another. The requirement that a solution concept should never dictate a choice that does not lead to a Nash equilibrium in any subgame is the crux of subgame-perfect equilibrium.

Definition 3.3 *A Nash equilibrium is a subgame-perfect equilibrium iff it induces an equilibrium in every subgame, that is, iff the Nash equilibrium strategies lead to Nash equilibria in every subgame.*

In the above example $(0,0)$ is the only subgame-perfect equilibrium. As the dominance-solvable equilibrium, the subgame-perfect equilibrium is a refinement of Nash equilibrium. It coincides with some Nash equilibrium, but does not necessarily accept all of them as the preceding example demonstrates. It should be observed that in order to determine the subgame-perfect equilibria we have to know the extensive form of the game at hand, while this information is not needed for the determination of Nash equilibria.

All two-person extensive-form games can be transformed into matrix-form games, but the converse is not always true. The crucial difference between these two forms is that in extensive-form representation one is able to model the type of information that players have when making their choices at various stages of the game. So far we have considered only complete-information games which are defined as games where the players know the payoffs of each other. Let us use the abbreviation K for 'the payoffs of the game'. In complete-information games, each player knows K. He also knows that the others know that he knows K. Moreover, he knows that the others know that he knows that they know that he knows K. This chain of knowing can in complete-information games be extended indefinitely. This amounts to the requirement that the payoffs in complete-information games be common knowledge.

A complete-information game may or may not be of perfect information (or of perfect recall). In perfect-information games the players know at which exact node of the extensive form they are when their turn comes. In imperfect-information games, there are at least some choices that have to be made without knowing at which node of the game tree one is located. The uncertainty concerning the location in the game tree is expressed in game theory by saying that in perfect-information games each information set is a singleton. In imperfect-information games at least one information set consists of several decision nodes. Consider as an illustration perhaps the best-known two-person zero-sum game, the Bismarck Sea Battle. Its extensive form representation is given in Figure 3.4. Its players, labelled K and I, were Admiral Kenney of the US Navy and Admiral Imamura of the Imperial Navy of Japan. The scene was in the Pacific War theatre during the Second World War. More specifically, the problem for Admiral Imamura was how to send a naval contingent to its destination New Guinea with minimal losses. Basically two routes were available: northern (n) and southern (s). Admiral Kenney, on the other hand, had exactly the opposite interests with regard to the convoy, that is, he wanted to use the US Navy air combat forces to cause as much damage to the Japanese contingent as possible. His estimate of the payoffs accruing to him under various scenarios are indicated in the figure. The payoffs could be interpreted, for example, as the number of effective bombing sorties. Clearly, he was anxious to guess the Japanese route choice correctly, while the Japanese were eager to have him make

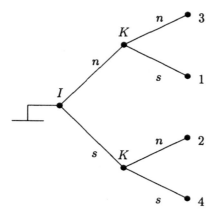

Figure 3.4: The Bismarck Sea Battle in Extensive Form

an incorrect guess.

At the time of deciding where to allocate the main part of his air power Kenney did not know the Japanese route plan. This fact has to be expressed in the extensive form by stating that the two nodes labelled U belong to the same information set. When two or more nodes are in the same information set, backwards induction does not work any more. Thus Zermelo's algorithm is not applicable in imperfect-information games. This is the main difference between backwards induction and subgame-perfect equilibrium.

Suppose, however, for the sake of argument, that Figure 3.4 is a perfect-information game. Then by applying Zermelo's algorithm the upper node 'U' can be replaced by $(3, -3)$ and the lower by $(4, -4)$. Thereafter, 'J' can be replaced by $(3, -3)$. Thus Zermelo's algorithm dictates that both players choose n. However, this conclusion is crucially dependent on the assumption that one player (USA) knows the choice of the other player.

In transforming an extensive-form game into matrix form, one starts from all information sets. A strategy of a player in the matrix-form representation consists of instructions concerning choices to be made in the player's every information set. In transforming Figure 3.4 into matrix form (Table 3.8) one observes first that there is only one information set for both players, one of which, namely Japan's, is a singleton. USA's information set, on the other hand, consists of two nodes, one following Japan's choice of n and the other following Japan's choice of s. By saying that these two nodes belong to the same information set we imply that, when making his choice, the US commander does not know the choice

Imamura Kenney	North	South
North	3	2
South	1	4

Table 3.8: The Bismarck Sea Battle

of the Japanese commander.

3.6   BELIEFS AND EQUILIBRIA

If both players of a two-person game have dominant strategies, it is natural to predict that those strategies will be chosen. For that prediction we need to assume that the players know their own payoffs. Knowledge of the other player's payoff would perhaps enhance the predictive capability of the dominance criterion, but is not really needed. In particular, one player does not need to know that his opponent has a dominant strategy and is likely to choose it. It is sufficient for him to know that, regardless of what the other player chooses and, thus, regardless of how rational he is, there is a unique best choice. Now, the luxury of not having to assume very much about the other player is absent in other solution concepts. They are all in varying degrees dependent on assumptions concerning the opponent's rationality.

If only one player has a dominant strategy and the other player a unique best reply to it, one would be tempted to say that the outcome is equally obvious as in the case where both players have dominant strategies. But this is not true. The player without a dominant strategy has to assume that his opponent is rational in the minimal sense of choosing a dominant strategy when he has one. In a pure-strategy Nash equilibrium where neither player has a dominant strategy the assumptions about the other player's rationality are even more important. We shall encounter games with several non-identical Nash equilibria in the following. In those situations the players may have opposed interests with regard to which a Nash equilibrium, if any, will be 'chosen'. Similarly, in the context of dominance-solvable equilibria, the assumptions concerning other player's rationality are crucial. We simply cannot predict that a dominance-solvable outcome will emerge in the game unless we assume that the players are rational in the sense of discarding dominated strategies and that they know that their opponents will act likewise. Since the Nash equilibrium plays such a crucial role in subgame-perfect equilibria,

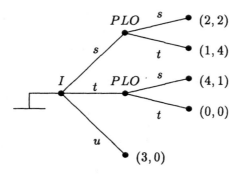

Figure 3.5: Israel–PLO Game

these considerations apply to these solution concepts as well. In addition to the more or less implicit assumptions about the other player's rationality made in these solution concepts, there are game-theoretic predictions that make explicit use of beliefs concerning the other player's behaviour. One of the solutions is that of perfect Bayesian equilibrium. Let us consider a highly stylized game model of the beginning of the peace process that started in the mid-1990s after the secret talks in Oslo (see Figure 3.5). There are two players, Israel and the Palestinian Liberation Organization (PLO). The issue is whether a negotiation process should be started that could eventually lead to a stable and harmonious coexistence of Israel and its neighbours. In our model Israel (I, in the figure) makes the first choice out of three options: to stay away from the negotiation table (u), to enter the negotiations with an inflexible stand on the issue of the Palestinian state, land for peace and so on (t), and to enter the negotiations with soft objectives (s). In this game we assume that the two nodes labelled PLO belong to the same information set, that is, the Palestinians don't know whether they are faced with a tough or soft Israel. The final branch leaves give the payoff pairs, the first element being the payoff to Israel and the second that to the PLO.

By going it alone, possibly using military means, Israel gets 3 and the Palestinians 0. The best outcome for Israel is that it assumes a tough stand and the PLO gives in. For the Palestinians this is better than an all-out conflict which Israel's u-choice would possibly bring about, but it is not a good outcome. The best the Palestinians can hope for in this game is 4, ensuing from their assuming an uncompromising negotiation stance, while Israel gives in. If both parties assume soft negotiation strategies the outcome is $(2, 2)$, a satisfactory outcome, but not best for either one. This outcome could be interpreted as something akin to the present (autumn 1997) situation in which it seems that (presumably) third parties are occasionally destabilizing the situation.

The concept of perfect Bayesian equilibrium is a specific application of the idea that underlies the Nash equilibrium, namely that the strategies of players should be the best responses to those of other players. But in addition this concept utilizes the beliefs that the players have with regard to each other's choices and imposes the requirement that the strategies are, indeed, the best ones, given the beliefs. In the Israel–PLO game, the PLO does not know at which node it is when called upon to make its choice, but it is plausible to assume that it can make more or less well-articulated guesses about Israel's negotiation posture. These can be modelled as a probability distribution over the two nodes in the PLO's information set.

Let us assume that the PLO thinks that with probability 1/4 it faces a soft Israel and with probability 3/4 a tough one. The expected utility of choosing s and t by the PLO can now be computed:

$$EU_{PLO}(s) = 1/4 \cdot 2 + 3/4 \cdot 1 = 5/4$$

$$EU_{PLO}(t) = 1/4 \cdot 4 + 3/4 \cdot 0 = 1.$$

Thus, choosing s leads to a higher expected payoff than choosing t, given this probability distribution. For all distributions that assign a smaller than 1/3 probability for 'soft', the PLO maximizes its expected utility by choosing 'soft'. In two-person games a pair consisting of beliefs and strategies is called sequentially rational if, given one player's beliefs, the other player's strategy and starting from any information set, the strategy of each player maximizes his expected utility in the remaining game. Consider, for example, the PLO's belief that if Israel enters the peace process negotiations, the probabilities of its 'soft' and 'tough' strategies are 1/2. So, the PLO's expected payoffs are:

$$EU_{PLO}(s) = 1/2 \cdot 2 + 1/2 \cdot 1 = 3/2$$

$$EU_{PLO}(t) = 1/2 \cdot 4 + 1/2 \cdot 0 = 2.$$

Thus, t would maximize the PLO's payoff under this probability assumption. Now, the strategies u of Israel and t of the PLO together with the PLO's belief that Israel chooses 'soft' and 'tough' each with probability 1/2 are sequentially rational. To see this, consider the possibility that Israel could change its mind about its choice. Could it improve upon the payoff 3 ensuing from its choice of u? The answer is no, provided that the PLO sticks to its strategy. To wit, the PLO's strategy would call for t and, hence, Israel's payoffs would be either 1 (if it chooses s) or 0 (if it chooses t). Thus Israel would not benefit from unilateral deviation. Could the PLO then benefit from it? The answer is again no since, given Israel's choice of u, the outcome is $(3,0)$ regardless of what the PLO chooses. Thus these strategies and the beliefs of the PLO constitute a sequentially rational outcome.

The perfect Bayesian equilibrium results from the application of Bayes's theorem to the probability assessments that the players make in the course of play. Bayes's theorem is usually applied in inferences from *a priori* to *a posteriori* probabilities of events. The former refer to guesses or hunches about the likelihood of events on the basis of prior knowledge or experience. The latter, in turn, are probabilities that the events occur, given some data that have been obtained through experiments or other kinds of observations. Consider a dichotomous event E. For example, let E denote the act of voting for party P in the previous elections. Suppose that the exit poll data of country X show that in the age group $25 - 40$, 5 per cent of voters voted for candidates of party P and that in other age groups the percentage of P's support was 6. Suppose, moreover, that 45 per cent of the electorate belongs to age group $25 - 40$. Let us denote the event of belonging to age group $25 - 40$ by A and the complement of this event, that is, the event of not belonging to this age group by $\bar{A}$. Thus,

$$P(E \mid A) = 0.05.$$

Suppose we want to find out the probability that a randomly chosen supporter of P belongs to age group $25 - 40$. That is, what is the probability $P(A \mid E)$ or what is the proportion of party P's total support that comes from voters in age group $25 - 40$? Bayes's theorem enables us to compute this probability:

$$P(A \mid E) = \frac{P(A)P(E \mid A)}{(P(A)P(E \mid A) + P(\bar{A}) - P(E \mid \bar{A})}.$$

Substituting the probability values we get:

$$P(A \mid E) = \frac{0.45 \cdot 0.05}{0.45 \cdot 0.05) + 0.55 \cdot 0.06} = 0.41.$$

The above process of converting the *a priori* probabilities into *a posteriori* ones is called Bayesian updating. It plays a role in the computation of perfect Bayesian equilibria. The perfect Bayesian equilibrium consists of beliefs and strategies so that the strategies are sequentially rational, given the beliefs, and the beliefs are updated using Bayes's theorem. In particular, if a player makes a choice that would be assigned a zero probability by the other player in a putative equilibrium, then that combination of strategies and beliefs cannot be a perfect Bayesian equilibrium.

In the preceding we have outlined some basic solution concepts in the theory of games. Our list is by no means exhaustive, but an attempt has been made to include those concepts that would be most pertinent in the design and evaluation of institutions. We shall now turn to some special types of games which have been used in modelling problems that have provided motivation for the design of institutions. Many of these problems relate to collective action and the construction of organizations to overcome free riding. Others pertain to coordination of activities to achieve common goals.

Further reading: Morrow (1994); Rubinstein (1982); Selten (1975).

3.7   BASIC ORDINAL 2 × 2 GAMES

The two-person constant-sum games are pretty thoroughly understood. The two-person non-constant sum games constitute a considerably more heterogeneous set. Even the most elementary ones, namely of the 2 × 2 variety, include a large set of strategically different games. Considering only the games with ordinal scale payoffs, that is, games in which the players have a complete and strict preference relation over the outcomes, there are 78 strategically different games. These do not allow for ties between outcomes (Rapoport and Guyer 1966). In other words, no two outcomes can be regarded as equally good by any player. The games are different in the sense of confronting the players with non-identical strategic uncertainties. In other words, the games cannot be reduced to each other simply by permuting the rows or columns of the matrices.

We shall focus on only a few special types of the 2 × 2 games. Our choice is dictated by relevance to institutional design. We start with by far the best-known game, namely the Prisoner's Dilemma.

3.7.1   Prisoner's Dilemma

Many, if not most, great ideas in the history of science have emerged without a recognition of their eventual relevance for scientific progress. This is certainly true of the Prisoner's Dilemma (PD for short) which, from a casual conversation topic grew into a huge branch of multidisciplinary research generating literally thousands of articles in scientific publications. The story that the game models has two persons arrested as suspects for a criminal act. The evidence for their conviction is in-

Suspect 2 Suspect 1	C	D
C	3,3	1,4
D	4,1	2,2

Table 3.9: Prisoner's Dilemma

sufficient. So the authorities try to get either one of the suspects to confess. The suspects are interrogated separately. They are encouraged to confess by a promise that the first one to confess is immediately given his freedom and a small reward for assistance to the authorities. If both confess, they are freed, but have to pay a fine. If neither confesses, they are both released because of the lack of evidence. If one confesses, the other one gets the maximum penalty, a short jail sentence. The situation can be modelled as a game played by the two suspects. The payoff matrix is shown in Table 3.9.

The payoffs are ordinal utilities. The choice C can be interpreted as a cooperative choice which in this context means that the player does not confess, while the choice D is a non-cooperative (or competitive) one, meaning that the player confesses. If both choose C, the outcome is next to the best one for both players (both are set free). The higher payoff 4 (freedom plus the reward) for one player is obtainable only at the cost of the other one who gets his lowest payoff (maximum penalty).

Let us define the binary relation of Pareto domination, denoted by D_P, over the set of outcomes X so that an outcome x (weakly) Pareto-dominates an outcome y, in symbols: xP_Dy, iff x brings at least as high a payoff to every player and to at least one player a strictly higher payoff than y. Notice that Pareto domination is a relation over set X of outcomes, whereas the relation of domination we discussed earlier is a relation between choices or strategies. An outcome is Pareto-dominated iff there is another outcome in X that Pareto-dominates it. Otherwise, it is Pareto-undominated.

The crucial features of PD can now be stated:

- D is the dominant strategy for each player.
- $(2, 2)$, resulting from dominant choices, is the only Nash equilibrium.
- There is an outcome, namely $(3, 3)$, that Pareto-dominates $(2, 2)$.
- The Pareto-undominated outcomes are: $(4, 1)$, $(3, 3)$, $(1, 4)$.
- Maximin-choices lead to the $(2, 2)$ outcome.

The unprecedented interest in PD can be explained by the observation that the game seems to capture the often-felt conflict between individual and collective rationality. The D choice is individually rational in the strongest possible sense since it dominates the other choice. Yet the outcome resulting from individually rational choices seems collectively 'irrational' in the sense that it is Pareto-dominated by another outcome.

The 'dilemma' in PD is the fact that cooperation would be beneficial for both players (up to a point), but it is unlikely since the achievement of the benefits of cooperation requires coordinated action of both players. The dilemma appears only when PD is understood as a non-cooperative game. A game is cooperative iff the players can make binding, that is, enforceable, commitments concerning their choices before the game is actually played. If in PD it were possible for the players to make an enforceable contract about the choices to be made, then the $(3, 3)$ outcome would undoubtedly emerge. But in order for an enforceable contract or some other type of binding commitment to be possible, there has to be a third party enforcing the contracts or otherwise securing that the commitments are really binding. To turn the, by definition non-cooperative, PD into a cooperative game is one way of 'solving' the dilemma. We shall discuss several others later on. For now, it is sufficient to notice that establishing a third party to enforce the contracts clearly means that an institution is designed to solve the problem of two-party interaction.

That PD-type settings require institutional means for the achievement of collective goals is also emphasized in the literature that uses PD in modelling the emergence of the state. In a hypothetical state-less original position the players are bound to choose individually rational strategies, whereupon the outcomes tend to be suboptimal. The suboptimality is to be understood as Pareto suboptimality. A state of affairs is Pareto optimal iff it is impossible to improve upon one individual's welfare without deteriorating at least one other individual's welfare. As was just pointed out, in PD the Pareto-undominated or Pareto-optimal outcomes do not include the individually rational $(2, 2)$ outcome. Hence, so the argument goes, the state is required to turn the non-cooperative PD game into a cooperative one.

Another way in which a third party can influence the outcomes in PD is 'quick and dirty', namely by changing the payoffs using rewards and penalties. Suppose that the third party wants to make the cooperative C choice more attractive to the players. Then it can simply add a bonus b to those payoffs that accrue to a player whose choice is C and/or cut down by p the payoffs to players choosing D. The ensuing payoff matrix would then look like Table 3.10.

By increasing b and p one eventually reaches a situation where PD

Column Row	C	D
C	3+b,3+b	1+b,4−p
D	4−p,1+b	2−p,2−p

Table 3.10: Modified Prisoner's Dilemma

ceases to be PD and C becomes dominant. This process takes place in many conflict settlement negotiations where the third parties, for example the international organizations, step in to encourage peaceful means of resolving PD-like conflicts through economic aid packages and so on.

Turning a PD into another game either by changing the payoffs or by making the game cooperative are 'solutions' to PD in the sense that they make the collectively desirable outcome also individually attractive. It should, however, be emphasized that in the strict game-theoretical sense PD has a solution and it is the non-cooperative outcome. The other solutions to PD to be discussed later on are attempts to render the *prima facie* irrational C choice rational and thereby achieve the $(3,3)$ outcome as a concequence of individually rational choices.

Although the discrepancy between individual and collective rationality is by definition always present in PD, there are factors which can exacerbate or ameliorate this discrepancy. The following list contains some intuitively relevant factors or parameters (Rapoport and Chammah 1965):

1. *Temptation*: this parameter is measured by the difference between best and next-best payoffs. Intuitively, the larger the temptation, the more it benefits a player to choose D or defect once he has allured his opponent to the belief that he will choose C. By increasing (decreasing, respectively) temptation, one would expect to be able to increase (decrease) the probability of the non-cooperative choice.

2. *Risk*: this is the difference between the payoffs denoted by 2 and 1 in Table 3.9. The payoff 2 is the value of PD, that is, the highest payoff that a player can unilaterally secure to himself. Thus by cooperating he takes the risk that his opponent does not cooperate, whereupon he gets his lowest payoff. Thus the larger the risk, the less incentive there would seem to be for cooperation.

3. *Gain*: this is the difference between the payoffs denoted by 3 and 2. This measures the benefit of cooperation over what could be received through individual rationality alone. Surely one way of making cooperation more likely would seem to be to increase gain.

The above parameters along with a large number of others have been

driver 2 driver 1	C	D
C	3,3	2,4
D	4,2	1,1

Table 3.11: Chicken

used in PD experiments, the earliest ones of which were conducted more than 30 years ago (Rapoport and Chammah 1965). By varying these parameters one gets a variety of PD games of varying incentives for cooperation and defection.

PD has often been regarded as the model of the collective action or collective goods provision problem. In particular, the phenomenon of free-riding finds its game-theoretic background in PD. If a benefit can be received without contributing to its costs, why contribute? Yet if everybody thinks this way, there will be no collective benefit. Under certain conditions this is, indeed, a PD situation. It would, however, be misleading to think that all collective action or collective goods provision problems are PD games. Sometimes the following game is a more appropriate model.

3.7.2   Chicken

The game of Chicken is derived from the Prisoner's Dilemma by interchanging payoffs 1 and 2. The anecdote underlying the name Chicken is from a face-saving contest of two teenagers. They are driving towards each other on a collision course at high speed. The first one to swerve is the loser (Chicken), his opponent the winner. If both swerve simultaneously, there is a tie and neither loses face. If neither one swerves, there will be a collision which brings the player the lowest payoff (loss of life or severe injury). Denoting the choice 'swerve' by C and 'not swerve' by D, we get the payoff matrix in ordinal utilities (Table 3.11).

The crucial characteristics of Chicken are the following:

- No dominant strategies for either player.
- Two pure strategy Nash equilibria: (4,2) and (2,4).
- Maximin-choices lead to a (3,3) outcome.

Although Chicken as well as PD is a 'mixed motive' game in the sense that the players have incentives for cooperative and non-cooperative choices, the prospect of ending up in the cooperative (3,3) outcome is essentially more likely in Chicken than in PD. One reason for this is that both players have an incentive to stay away from the D choice since

Column Row	C	D
C	4,4	1,3
D	3,1	2,2

Table 3.12: Assurance Game

the worst outcome is associated with this choice. Another reason is the fact that there are no dominant strategies in Chicken. The prediction of what will be the outcome is thus less straightforward than in PD. The Nash equilibrium concept is not very helpful since, considering the pure strategies only, there are now two very different Nash equilibrium outcomes; each brings one player the best payoff and the next to worst to the other. If, for a moment, we regard the payoffs as ratio scale utilities, then even a third Nash equilibrium can be found, namely one in which both players choose C with probability $1/2$ and D with probability $1/2$. As always, its interpretation in a one-shot game is difficult.

3.7.3 Assurance Game

An Assurance Game results if one interchanges the payoffs 4 and 3 in PD. As a consequence, the players receive their largest payoffs in the same outcome, namely one that results from C choices of both players (see Table 3.12).

The essential features of the Assurance Game are the following:

- There are no dominant choices.
- There are two Nash equilibria: $(4,4)$ and $(2,2)$.
- One outcome, namely $(4,4)$, Pareto dominates all others.
- Maximin choices lead to the Pareto-suboptimal Nash equilibrium $(2,2)$.

Although everything seems to point to the Pareto-optimal Nash equilibrium as the most likely outcome in the Assurance Game, it is worth noticing that the second largest payoff is obtained by a player who chooses D or defects, while the other player chooses C. The latter player then gets his worst payoff. Thus by choosing C unilaterally a player takes a risk that his opponent is not rational or does not see the game as an Assurance Game. The latter possibility means that the game is of incomplete information.

The Assurance Game is sometimes used in modelling the adoption of certain common practices or symbols to facilitate human interaction. An important example is currency as a symbol and carrier of value (Lagerspetz 1984). It is to the benefit of both players that a common standard or rule is adopted, but there still remains the problem of making its acceptance known to the other player so that he, in turn, may accept it without risk. Various kinds of institutions exist just to solve this type of coordination problem (traffic rules, alphabet systems, maps, and so on).

Thus, all three types of ordinal 2 × 2 games we have discussed above involve problems that are solved by institutions. In other words, institutions are sometimes designed to avoid the problems described in those games such as ending up with Pareto suboptimal outcomes. It is, however, important to observe that behaviour within an institutional framework can also be regarded as a game. Thus institutions are also redesigned to ameliorate problems related to the previous ones. A natural way to proceed in designing an institution within which rational behaviour results in desired equilibria is to modify the payoff structure of the game played within the existing institution. Methods used in the modification usually include rewards and penalties. We shall discuss some problems related to this seemingly obvious way to bring about desired outcomes as equilibria.

Further reading: Lagerspetz (1984); Sen (1967).

3.8   THE IMPORTANCE OF RIGHT MODALITY

George Tsebelis introduces the notion of the Robinson Crusoe fallacy to describe an inference that erroneously treats a strategic environment as a passive state of nature (Tsebelis 1989). In other words, one commits this fallacy when one confuses risk with strategic environment. In the former one is entitled to assign objective probabilities to states of nature and compute expected utilities of choices given those probabilities. In a strategic environment there are always several actors whose interests are at stake. If the system requires that individuals behave in certain ways, for example stop smoking or start saving, it may introduce rewards for good behaviour and penalties for bad. The fallacy lurks in not realizing that those very rewards and penalties may affect the system's utilities as well. Tsbelis argues that this is often bound to happen in equilibrium. Lets us consider Tsebelis's example.

In an effort to limit speeding in certain areas the police may resort to radar controls positioned in those areas and fine the speeders caught. The conventional wisdom has it that higher fines and/or higher probability of being caught makes the driving public less likely to speed. The drivers' payoffs are those expressed in Table 3.13. The columns denote

state of nature public	police present	police not present
speed	a_1	b_1
not speed	c_1	d_1

Table 3.13: Traffic Control Problem

two states of nature: the presence and absence of a police patrol. It seems natural to assume that $c_1 > a_1$ and $b_1 > d_1$.

In other words, if the patrol is present it is better for the driver not to speed, while speeding gives a higher payoff if the police is not present (assuming that the drivers enjoy fast driving or are in such a hurry that speeding is the only way to get to their destinations in time). The payoffs express total utility levels of the public. We denote the probability of police presence by p. Hence, an EU maximizer speeds iff

$$EU(speed) > EU(not\ speed)$$

or

$$[pa_1 + (1 - p)b_1] > [pc_1 + (1 - p)d_1].$$

Suppose that the probability p is constant. If one wants to reduce the number of speeders, one should reduce the left-hand side of the above inequality. The most realistic way of doing that is to reduce a_1. This, in turn, means increasing the penalty, for example the fines. If p is not constant, the other possibility (which does not exclude the first) is to increase p. These methods lie at the heart of many control policies.

The methods are, however, based on the assumption that the situation is a game against nature or a traditional decision problem in a passive environment rather than a genuine game between two players. The picture of the policy maker as a passive state of nature is, however, intuitively self-contradictory. After all, it is the policy-maker who uses the modelling devices in an effort to achieve some specific goals. So, Tsebelis suggests that the situation ought to be modelled as a game of two players, the public and the police (see Table 3.14).

It is plausible to assume that

$$c_1 > a_1; b_1 > d_1$$

and

$$a_2 > b_2; d_2 > c_2.$$

police public	enforce	not enforce
speed	a_1, a_2	b_1, b_2
not speed	c_1, c_2	d_1, d_2

Table 3.14: Traffic Control Game

Suppose now that the police and the public can resort to mixed strategies. In the case of the public these have a relatively straightforward interpretation: strategy (*speed, p; not speed,* $1-p$) means that $100p$ per cent of the drivers speed. The strategy (*enforce, q; not enforce,* $1 - q$) of the police could be interpreted purely probabilistically, that is, as indicating the probability of the patrol being present in the area, or as the percentage of time that the patrol is present.

The computation of the mixed strategy Nash equilibria yields:

for public: $p^* = (d_2 - c_2)/(a_2 - b_2 + d_2 - c_2)$
for police: $q^* = (b_1 - d_1)/(b_1 - d_1 + c_1 - a_1)$.

These define the unique equilibrium strategies of the game. Now, the public controls the value of p, while the police determines the value of q. A look at the previous formulas of the equilibrium values reveals that the utility of speeding and getting caught, a_1, does not appear at all in the equilibrium value of p. In other words, in the equilibrium the increase or decrease of the penalties does not affect the public's probability of speeding. What it does affect is the equilibrium value of q, that is, the probability of enforcement. The larger the fine, the larger is the difference $c_1 - a_1$ and, thus, the smaller q^*, that is, the larger the penalty, the smaller the probability that the police enforces the law at the equilibrium. Thus the increase in penalties has no effect on the behaviour of the public, but paradoxically enough seems to reduce the probability that the law is being enforced by the police. This clearly contradicts the implication of the decision-theoretic analysis. Intuitively it would seem that the game-theoretic approach is more appropriate than the decision-theoretic one. After all, in the former the fact that the police also have interests and preferences is explicitly taken into consideration.

The above example suggests that the decision modality one deals with is an important consideration. It pertains to a very wide range of public policy problems. Whenever an agent tries to influence another one to do something or to refrain from doing something, the issue of rewards and punishments emerges. Tsebelis has applied the above considerations to an analysis of the reasons for the failures of economic sanctions to bring about the intended behavioural changes in the target countries and to the compliance of industries with government regulation (Tsebelis 1990; 1991). The point in these examples is to show that once both the policy maker or designer and the 'target' of the policy are modelled as strategic actors, the equilibrium behaviours have completely different, indeed, contradictory determinants than in cases where the policy maker is modelled as a passive state of nature.

Despite their in principle profound implications for institutional and policy design, one should observe, however, that the mixed-strategy Nash equilibrium strategies have very weak predictive value. In other words, the fact the equilibrium strategies can be computed gives very little guarantee that they would be 'found' by the players (Bianco et al. 1990). All that this equilibrium concept guarantees is that once one of the players has found his equilibrium strategy, the other one cannot benefit from not resorting to his equilibrium strategy. Thus, once the players have hit the equilibrium, they are likely to stick to their strategies. But to find his strategy a player has to know the payoffs of his opponent. Hence the assumption of complete information is crucial here. Furthermore, the payoffs have to be cardinal utilities for the mixed strategy computations to yield meaningful results.

Despite these practical reservations, the lesson taught by the above example is very important for all control policy. Increasing penalties or rewards may not necessarily lead to the intended behaviour in equilibrium. Yet it is precisely at the equilibrium behaviour that the policies are typically aimed. For example, in changing the tax code one strives at changing economic behaviour after the initial adjustments, that is, in equilibrium. Certainly the anticipated adjustments may sometimes also be taken into account, but especially in designing institutions or long-term policies the equilibrium state is quite crucial. Thus Tsebelis's examples show that the proper decision modality needs to be given serious consideration.

Further reading: Tsebelis (1989); Bianco et al. (1990).

3.9 NON-MYOPIC EQUILIBRIA

In the discussion on Tsebelis's model it has been pointed out that the model is too static to give an accurate picture of the decision-making situation (Bianco et al. 1990). Indeed, the Nash equilibrium outcome is based on calculations that the players may have in mind once the equilibrium has been reached. Suppose, however, that the game begins in a fixed status quo outcome of the payoff matrix and the players ask themselves if they could improve upon their position by changing their strategy, taking into account the other players' likely response. This is the approach to the equilibrium problematique adopted by the theory of moves (TOM for short)(Brams 1994). This theory has mainly been applied to explain conflict behaviour, but its principles can also shed light on problems of institutional design. To wit, TOM is a tool for predicting non-myopic behaviour. Since the design of institutions requires knowledge of principles that guide rational actors, TOM is a useful instrument to the extent that it has succeeded in explaining otherwise unexplained and yet intuitively rational behaviour.

TOM focuses on 2 × 2 ordinal games. In contrast to the games we have discussed thus far, TOM assumes that the payoffs are not received by the players until an equilibrium outcome is reached. Starting from any given outcome in the payoff matrix the players are assumed to make their choices in strictly alternating fashion until the final outcome — to be discussed in a moment — has been reached. Row is allowed to move the outcome up or down in the given column, while Column is allowed to move the outcome horizontally staying in the same row. The rules of TOM include only moves in a clockwise or counterclockwise direction away from the starting outcome. From the final outcome one then retreats until the starting outcome, using at each stage the principles of backwards induction.

With regard to the definition of the final outcome, there are several versions of TOM. In the earlier one (Brams and Wittman 1981), the final outcome or cell of the matrix is reached when that player whose turn it is to move (moving player, for short) would get the payoff 4, that is, his maximum, in that cell. So he has absolutely no incentive to move. Let us suppose that this player is Row. Once such a final outcome is reached one asks whether this outcome is such that Column would prefer it to the payoff that he (Column) would get in the immediately preceding cell. If Column prefers the preceding cell, then that outcome survives the comparison with the final outcome. Otherwise the final outcome survives this comparison. One then asks whether the outcome that survives would be preferred by Row to the payoff received in the immediately preceding cell. Proceeding in this fashion one comes to the initial outcome and can determine whether it survives the comparison with the outcome that has survived the previous comparison. If the outcome that survives the last comparison is the initial one, then the player who started the chain of alternating moves has not gained anything. Rather, the process which he originated leads to an outcome that is worse than the one he started from. If this applies for both players, then the original outcome is a non-myopic equilibrium. Let us state the definition somewhat more precisely.

Definition 3.4 *An outcome* (a_1, a_2) *is a non-myopic equilibrium for*

Row iff the process of alternating moves that Row starts in the 2×2 game results in final outcome (x_1, x_2) so that a_1 is strictly preferred to x_1 by Row. An outcome is a non-myopic equilibrium iff it is a non-myopic equilibrium for both Row and Column.

This is the early version of TOM. The more recent version differs from this in the definition of the final outcome as well as in the explicit definition of the rules of play (Brams 1994; Brams and Mattli 1993). The main rules are:

1. The play starts at a fixed cell of the payoff matrix, that is the initial state.

2. The play proceeds through strictly alternating moves, that is, one player moves after the other.

3. The moves continue until the final state is reached. That is a state where the player whose turn it is chooses not to move. In practice, the final state is the outcome that survives the fourth comparison which is between the initial state and the outcome that immediately precedes it.

4. A player will not move from the initial outcome if the process ends in a less preferred final outcome or returns to the initial state.

5. The game is of complete information.

Consider as an example the PD game of Table 3.9 on page 69 and assume that $(2, 2)$ is the status quo. Let us first find the final outcome starting with Row's move from $(2, 2)$ to $(1, 4)$. Then Column moves from $(1, 4)$ to $(3, 3)$, whereupon Row again moves to $(4, 1)$. Column now has the choice of either moving back to 'square one', that is to the initial state $(2, 2)$ or stopping at $(4, 1)$. Since the former outcome is preferred to the latter by Column, we can assume that the final outcome, that is, one that survives the first comparison, is $(2, 2)$. Now this outcome is compared with $(3, 3)$ using Row's preferences. Clearly $(3, 3)$ wins. Then $(3, 3)$ is confronted with $(1, 4)$ using Column's preferences, whereupon the latter is found preferable. Finally, $(1, 4)$, the surviving state thus far, is compared with the *status quo* $(2, 2)$ using Row's preferences, whereupon the initial state is found preferable. Thus we may conclude that Row, anticipating this chain of moves and countermoves, never makes the first move.

Now, a similar reasoning applies to Column when $(2, 2)$ is considered the initial state. Thus $(2, 2)$ is a non-myopic equilibrium if it is the initial state. Suppose, however, that the initial state is $(3, 3)$. If Row makes the first move, the final outcome is $(1, 4)$ since Column prefers it to $(3, 3)$. $(1, 4)$ does not, however, survive the comparison with $(2, 2)$ when Row's preferences are decisive. Rather, $(2, 2)$ is compared with $(4, 1)$ using Column's preferences, with the result that $(2, 2)$ survives. Finally, Row compares the initial state $(3, 3)$ with $(2, 2)$ and observes that $(3, 3)$

survives. Thus $(3,3)$ is also a non-myopic equilibrium from Row's point of view. By a similar argument one sees that $(3,3)$ is also a non-myopic equilibrium for Column. Thus we have two non-myopic equilibria in PD, each with a different initial state. The surviving outcome starting from $(1,4)$ with Row moving first is $(2,2)$ and $(1,4)$ with Column moving first. Thus the situation is indeterminate unless other rules for moves are given. Brams suggests that if an outcome is a non-myopic equilibrium for one player but not for the other, it is the latter's incentive to move that is decisive. On the basis of this rule, then, $(1,4)$ is not a non-myopic equilibrium. Similarly, it can be seen that $(4,1)$ is not a non-myopic equilibrium.

The principles of TOM can be extended to the other 2×2 ordinal games. In Chicken it turns out that there is only one non-myopic equilibrium, namely $(3,3)$. In other words, if $(3,3)$ is the status quo, then it remains so in the sense of non-myopic equilibrium. No other outcome has this kind of stability in the Chicken game. In the Assurance Game there is also one non-myopic equilibrium, namely $(4,4)$. Thus, as the case of Chicken demonstrates, the non-myopic equilibrium is not a refinement of a Nash equilibrium.

Anticipation of other players' choices together with the initial cooperation, are explanations derived from TOM for the fact that sometimes players in PD situations choose to cooperate rather than defect. Since PD-like situations are quite common in human affairs, a number of explanations for individually irrational cooperative behaviour have been suggested in the literature. Some of them will be reviewed in the following. Uncovering reasons for cooperative behaviour in PDs can be useful to an understanding of why certain types of dilemmas need institutional resolutions, while others result in the desired outcomes by themselves. In the following we shall give a brief overview of some attempts to make the cooperative $(3,3)$ outcome an individually rational solution to the dilemma.

Further reading: Brams (1994).

3.10 ANALYTIC SOLUTIONS TO THE PRISONER'S DILEMMA

In this section we shall focus on work that has been done in rendering the collectively 'rational' outcome also individually rational in the sense that individuals finding themselves in PD-like situations would find it sensible to resort to cooperative choice, whereupon the $(3,3)$ outcome would ensue. A straightforward way of doing this would be to invoke some

Column Row	C/C	C/D	D/C	D/D
C	3,3	3,3	1,4	1,4
D	4,1	2,2	4,1	**2,2**

Table 3.15: First PD Metagame

collectivistic considerations on the part of the players, but this would conflict with the overall individualistic − or at least reductionistic − programme underlying game theory. So, 'group spirit', 'collective ethos' and explanations like that will not do since they are not *explanantia*, that is, explaining entities, but rather *explananda*, things that call for an explanation.

3.10.1 Howard's metagame solution

The argument that Nigel Howard presented in the early 1970s to support the choice of cooperative strategy in PD was the following (Howard 1971). Suppose that, instead of choosing between C and D, the players − perhaps anticipating that they will see each other again later on or that the game will anyway be played several times by each player or out of a wish to act upon a principle rather than *ad hoc* − choose conditional strategies. Then it turns out that there exists a pair of conditional strategies that is a Nash equilibrium and that calls for a cooperative choice to be made by both players. Let us discuss this argument in some detail.

The conditional strategies are rules that make one's choice dependent on some event. In particular, let us assume that the players condition their choices on each other's behaviour. So one might consider a conditional strategy 'choose D if the opponent chooses C, choose C if he chooses D', or 'choose exactly as the other player does'. This latter strategy is the famous TIT-FOR-TAT (TFT for short) which some ten years after Howard's book was to become a major subject in the study of the emergence of norms and institutions. Adopting the notational convention that x/y means that x is chosen if the other player chooses C and y is chosen if he chooses D, these two strategies can be written as D/C and C/D, respectively.

Howard suggests that one constructs the first metagame, that is, a game which is derived from the original game (PD) by conditioning one player's strategies on the choices of the other player. Let us assume that Row's choices remain unchanged, but Column's strategies are conditional on Row's choices. Thus we get the payoff matrix shown in Table 3.15.

Column Row	C/C	C/D	D/C	D/D
$C/C/D/D$	3,3	**3,3**	4,1	2,2
$C/D/D/D$	3,3	2,2	4,1	2,2
$D/D/D/D$	4,1	2,2	4,1	2,2
$D/C/D/D$	4,1	**3,3**	4,1	2,2
$C/C/C/C$	3,3	3,3	1,4	1,4
$C/C/C/D$	3,3	3,3	1,4	2,2
$D/C/C/C$	4,1	3,3	1,4	1,4
$D/D/C/C$	4,1	2,2	1,4	1,4
$D/D/D/C$	4,1	2,2	4,1	1,4
$C/D/C/C$	3,3	2,2	1,4	1,4
$C/C/D/C$	3,3	3,3	4,1	1,4
$D/D/C/D$	4,1	2,2	1,4	2,2
$D/C/D/C$	4,1	3,3	4,1	1,4
$D/C/C/D$	4,1	3,3	1,4	2,2
$C/D/D/C$	3,3	2,2	4,1	1,4
$C/D/C/D$	3,3	2,2	1,4	2,2

Table 3.16: Second PD Metagame

Column has four conditional strategies or metastrategies since the two symbols C and D can replace x and, independently of this, the same two symbols can replace y in the above expression. The payoff matrix is formed by observing that once Row's choice is fixed, Column's choice is also known. It is simply the response to Row's choice. For example, the entry $(3,3)$ in the second column of the first row is the result of Row's choice of C and Column's choice of C/D since the latter calls for C in case the opponent chooses C.

A look at Table 3.15 reveals that no new equilibria emerge. Rather, $(2,2)$ at the intersection of D and D/D, calling for each player to choose D, remains the only Nash equilibrium. However, if one also makes Row's choices conditional on Column's strategies — or in this case metastrategies — we get new equilibria. Table 3.16 indicates the payoffs for the game in which both players' strategies are conditional on each other. Since Column has four metastrategies and Row can respond to each one of them in two ways (either C or D), Row has altogether $2^4 = 16$ different metastrategies. These are indicated as rows in the table. For example, $C/D/C/D$ is a metastrategy that responds with C to Column's C/C, with D to Column's C/D, with C to Column's D/C and with D to Column's D/D.

Payoffs are determined as follows. Consider the first column of the second row. It is at the intersection of Column's C/C and Row's $C/D/D/D$. The first element (C) in the latter strategy indicates which choice is made if the opponent chooses C/C, and since Column chooses C/C, the outcome is that Row chooses C, whereupon Column's meta-strategy calls for the choice of C as well. Thus $(3,3)$ ensues.

In the second metagame there are two Nash equilibria: one at the intersection of $D/C/D/D$ by Row and C/D by Column and one at the intersection of $C/C/D/D$ by Row and again C/D by Column. The interesting observation about these equilibrium strategies is that they both imply a choice of C, whereupon $(3,3)$ emerges as a Nash equilibrium outcome. In addition to these two $(3,3)$ equilibrium outcomes, there are no other equilibria in the second metagame.

Thus it appears as if the cooperative choice will be individually rational in PD if the players make norm-based choices, that is, consider the possibly unique PD game as an instance of similar situations in which various types of norms can be used. The norms they come up with at the Nash equilibrium express pure reciprocity on the part of the player whose choice is C/D and a cooperative attitude towards reciprocity on the part of the other player.

Howard argues that constructing higher-order metagames does not bring about new equilibria. Thus Table 3.15 and Table 3.16 reveal the whole story about metagame analysis. Yet one may wonder about the meaning of reciprocity in a one-shot PD. Since the players are assumed to make their choices not knowing the choice of each other, C/D remains a purely hypothetical choice, indicating what one would do if one knew the opponent's choice. Reciprocity as a norm guiding behaviour can only have meaning in situations involving several encounters of players, either in the context of repeating the same game (for example PD) or in some other game context. Indeed, TFT has been extensively discussed in the repeated context where its equilibrium properties have a more natural content. We shall return to repeated PDs later on.

Howard's early book put forward two new ideas that later turned out to be very important to an understanding of repeated games: the idea of norms or rules that the players observe in their interactions and that by themselves may be successful or unsuccessful, and the concept of reciprocity. Norms or rules can, on the one hand, be regarded as tools that enable the decision maker to decide a large number of choices in one policy decision, thereby releasing his computational resources for other uses. On the other hand, the very complexity involved in playing PD or some other game successfully in a long sequence of repetitions poses the problem of computational bounds, that is, how difficult are the tasks that decision makers can be expected to solve in an effort to play a game 'optimally'? From Howard's analysis onwards the focus of PD research moved towards sequential PDs and good long-term strategies in them, as well as to issues related to the limits of players' computational capabilities.

Further reading: Howard (1971).

3.10.2 Taylor's supergame solution

Michael Taylor took the logical next step in applying the TFT-strategy to iterated PD setting (Taylor 1976). His theoretical argument is directed against the relatively widespread view that the state or other higher-order authority is needed to solve collective action problems. He argues that this view is misleading in two respects. First, it assumes that collective action problems (for example restricting the exploitation of non-renewable natural resources or prevention of air and groundwater pollution) are necessarily or even typically PDs. In Taylor's opinion they are not. Second, it implies that sequences of PDs played over time would necessarily call for a higher authority to enforce the cooperative choices. This is not necessarily the case, either, says Taylor.

Taylor's technical focus is on strategies in so-called supergames which are games consisting of a sequence of elementary one-shot games. The latter can be, for example, PDs. A supergame strategy is a rule that assigns a choice in every constituent game. Taylor tries to find out what kind of supergame strategies are successful if the elementary games are all identical PDs. Not only are the PDs assumed to be identical, but they are also assumed to be played by the same players. Thus issues like reputation, memory and cheating become pertinent. Intuitively, norms of behaviour are more applicable in these kinds of settings than in one-shot games.

The supergame success depends on the payoffs in the constituent PDs. Since one needs to compare the payoffs accumulated over time, it has to be assumed that the payoffs are measurable on a cardinal scale. Specifically, we define the supergame payoff of a player as the sum of his payoffs in each constituent game weighted by a discount parameter a^t, where t denotes the number of time periods between the present and the time instant at which the game is played. Thus player i's supergame payoff, that is, the payoff he will receive in an infinite sequence of PDs, is $P_i = \Sigma_t a^t X_t$, with $t = 1, \ldots, \infty$. The discount parameter is assumed to be in the $(0, 1)$ interval. This guarantees the convergence of P_i. The discount parameter is a measure of the player's interest in payoffs that he receives later on in the game. The closer to zero the discount parameter, the less weight the player assigns to payoffs received after a few rounds. Conversely, the closer to unity the discount parameter, the more weight the long-term payoffs have in the player's deliberations.

Taylor considers a relatively small subset of possible supergame strategies. To wit, the following strategies are dealt with:

- C^∞: the strategy that dictates C in every constituent game.
- D^∞: the strategy that dictates D in every constituent game.
- A_k: the strategy that dictates C as long as the opponent has chosen C in the previous game. The first defection of the opponent is punished with k defections, whereupon C is resumed and continued until the second defection of the opponent. This eventuality is responded with $k+1$ defection after which C is again resumed, and so on.
- A_∞: the strategy dictates C until the opponent defects. After the defection, it dictates D for all remaining games.
- TFT: C is chosen in the first game. After that the choice is the same as that of the opponent in the previous game.
- B: D is chosen in the first game. After that the strategy is identical with TFT.

Within this restricted set of supergame strategies the results suggest that the only robust equilibrium strategy pair is (D^∞, D^∞). There are other equilibria, but these require additional assumptions. On the other hand, Taylor's analysis shows that the TFT pair is also an equilibrium, provided that the discount parameter is large enough. This observation was expanded a few years later into a theory of emergence of cooperation by Robert Axelrod. Before discussing that theory, let us take a look at another related approach to PD.

Further reading: Taylor (1976; Taylor 1987).

3.10.3 The good strategy

Steven Smale applies an approach stemming from the study of dynamical systems to PD with some interesting results. He first defines a criterion of performance for supergame strategies as the unweighted average of payoffs received in a sequence of PD games. This criterion requires that the payoffs be measurable on a cardinal scale. Figure 3.6 depicts a particular PD, namely one with payoffs $(3,3), (4,1), (2,2)$ and $(1,4)$. The payoffs are represented by the vertices of a quadrangle S. Since each constituent game in the supergame is a PD, its outcome is one of the vertices. Repeated play starting at time 1 yields a sequence $x_1, x_2, \ldots, x_i \in S$.

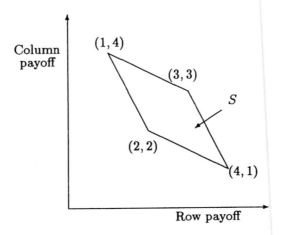

Figure 3.6: PD Payoffs in Two-Dimensional Space

Consider now the average payoffs, that is, define

$$q_T(x_1, \ldots, x_T) = \frac{1}{T} \sum_{i=1}^{T} x_i = X_T.$$

The set S consists of all points that can be obtained as probability mixtures of the corners of S. Thus we can focus our attention on the points in S and rest assured that no supergame strategy can give a player an average payoff which is not representable as a point in this set.

The payoff function is $\pi : A \rightarrow S$, where $A = \{C, D\} \times \{C, D\}$. In other words, each pair of choices is mapped into a point in S. Smale considers strategies that are maps $s : S \rightarrow A$, where $s = (s_1, s_2)$. This means that the strategies make the choice in the next PD round dependent on the average of payoffs received so far. Thus, for fixed s, the starting point, x, determines the entire evolution of the game. Thus, given an initial payoff distribution and a rule that assigns C or D to the average payoff obtained, one is able to trace what happens in the game in all subsequent periods.

The supergame can be represented as a dynamical system, that is, a system which at any given time is in some state and moves from one state to another according to rules typically represented as a state-transition function. These are mappings

$$\beta_T : S \rightarrow S$$

where S denotes the set of states of the system. In the PD system, S is the set of possible average payoff distributions for the players. Since the players are assumed to be interested in the average payoffs, we can specify β_T as follows:

$$\beta_T = \frac{Tx + \pi \otimes s(x)}{T+1}, \quad x \in S.$$

Here $\otimes$ denotes the operator product, that is, the result of applying two operators consecutively. In this case, $\pi \otimes s(x)$ denotes the result of first applying s to point x in S and then applying π to the result of the previous operation. More specifically, for given x, $s(x)$ gives the choice to be made by each player in the next round, whereupon $\pi \otimes s(x)$ indicates the outcome in the next round of PD.

In a dynamical system some states may be stationary. This means that for such a state x, if $x = x_1$, then $x_{T+1} = x_T$, for all values of $T = 1, \ldots$. Interpreting states as pairs of average payoff, a stationary state is an average payoff distribution between the two players that, once reached, will remain unchanged for the rest of the game.

A solution in this system is a pair (s, x), where $s : S \to A$ is a strategy and $x \in S$ is stationary with respect to s. In other words, when in a solution the strategies are so defined that the average payoffs remain the same for all continuations of the game. A solution is globally stable if all initial states eventually lead to it. More formally, consider a dynamical system defined by a transition function β_T ($T = 1, 2, \ldots$) so that $x_{T+1} = \beta_T(x_T)$. In this system (s, x) is globally stable if for any $x_1 \in S : x_T \to x$, when $T \to \infty$. In other words, a globally stable solution will eventually be reached starting from any initial state (payoff pair). It is, thus, a very strong prediction of what will happen − at least in the long-run − in the system under study.

As was pointed out above, strategies are rules that indicate whether choice C or D is to be made when the average payoffs (up to the present) are known. Thus, each strategy has two arguments: the average payoff to player 1 and the average payoff to player 2. In Smale's terminology a strategy s_1 for Row is a good strategy if:
1. $s_1(\alpha, \beta) = D$ if $\alpha < 2$ (because by always choosing D Row would guarantee that $\alpha \geq 2$), that is, a good strategy dictates the choice of D if one's own average payoff is less than 2.
2. $s_1(\alpha, \beta) = D$ if $\beta > 3$ (because $\beta > 3$ means that Column has been exploiting Row at least part of the time).
3. $s_1(\alpha, \beta) = C$ if $\beta \leq \alpha$ and there is some open set U in S which contains the segment $\{(k, k) \in S | 2 < k < 3\}$ on which $s_1 = C$.

The first condition states that one should choose D if one's average payoff falls below 2. The second says that if the opponent has been doing on average better than he would if both players cooperated, then one should resort to D. The third condition says roughly that if the opponent is doing no better than oneself, then C should be chosen. The good strategy is obviously not the same as a predominantly cooperative strategy. The first two conditions are precautions against exploitation. On the other hand, the third condition is somewhat biased towards choice C since C should be chosen whenever the opponent is doing worse or at most equally well as oneself.

Smale proves two important theorems. The first one is the following. Let Row play a good strategy. Then

$$\lim \inf x_T^{Row} \geq 2.$$

In other words, by playing a good strategy Row can rest assured that the limiting average payoff is not less than his security level 2. Moreover, $\lim \sup x_T^{Col} \leq 2$ (that is, Column can do no better than by playing a good strategy, too). If both play a good strategy $(s_1, s_2) = s^0$, then $x' = (3, 3)$ is a solution and is the unique x' such that (s^0, x') is a solution. $(s^0, (3, 3))$ is globally stable.

In other words, by playing a good strategy a player can guarantee himself a payoff that is no worse than his security level, that is, the payoff that he is able unilaterally to guarantee himself. When one player plays a good strategy, the other cannot benefit from not playing one too. Moreover, good strategies by both players secure the global stability of the $(3, 3)$ outcome.

However, as Smale's second theorem shows, only a slight modification in the third condition is needed to make $(2, 2)$ the unique globally stable outcome. This modification is the following: $s_1(\alpha, \beta) = D$ if $\alpha \leq \beta$ and there is some open set $U \in S$ which contains the segment $\{(k, k) \in S | 2 < k < 3\}$ on which $s_1 = D$.

Once the third condition is replaced with this one, $(2, 2)$ emerges as the sole globally stable outcome. Intuitively, the modification means that when both players have the same average payoff, then the strategy calls for D, while a good strategy would call for C. When $\beta > \alpha$ the third condition and its modification are compatible in the sense that the former dictates nothing, while the latter calls for D. Similarly, when $\beta < \alpha$ a good strategy calls for C, while the modification imposes no restriction on the choice.

Smale's results show that a good strategy is successful in PD environments where the opponent's strategy is not necessarily a good one. In particular, nothing is gained by not choosing a good strategy if the opponent is choosing a good one. Thus, a pair of good strategies forms a Nash equilibrium. The third condition in the definition of a good strategy and its slight modification show, however, that the stability of the cooperative outcome is dependent on exact observations of the other player's average payoffs. In many contexts this is an unreasonably stringent assumption. The dynamical systems approach shows, however, that repetition of PD opens new vistas for the possibility of cooperative outcomes.

Smale's theorems show that once good strategies have been adopted by the players, there is no reason for them to refrain from using them indefinitely. As such the theorems do not, however, give any reason for the players to adopt good strategies in the first place. In other words, the theorems do not predict that good strategies will eventually be 'discovered' by players and then used for the rest of the supergame.

It is noteworthy that a good strategy is a variant of TFT. The latter starts with cooperation and then proceeds to choose whatever the opponent has chosen in the previous game. A good strategy is also based on monitoring the opponent's previous choices. In fact, this is done both directly (condition 2 of the definition) and indirectly (condition 1 of the definition). In TFT the monitoring takes place immediately after each round of PD, whereas a good strategy is based on observing the average payoffs. These differences and similarities of the dynamical systems approach and TFT will become evident once we have discussed TFT in more detail.

Further reading: Smale (1980); Rubinstein (1986).

3.11 PRISONER'S DILEMMA AND EVOLUTIONARY STABILITY

In the late 1970s Robert Axelrod organized a tournament of PD-playing computer programs. The aim was to find out which one would be most successful in an artificial evolutionary environment consisting of other programs that had been submitted to the tournament (Axelrod 1980a; 1980b). The invitation to enter the contest was sent to a group of well-known game theorists. The instructions for contestants indicated the setting: the programs would be playing against each other and against a program which simply chooses C and D at random. The success would be determined by the accumulated payoffs in a long sequence of PD games. The length of the sequence was not indicated.

Fourteen programs were competing in the first tournament. One of them was TIT-FOR-TAT submitted by Anatol Rapoport. A round-robin tournament consisting of each program playing 200 rounds of PD against each other program was conducted. The matrix of the cardinal payoffs used in the tournament is given in Table 3.17.

	C	D
C	3,3	0,5
D	5,0	1,1

Table 3.17: The Payoff Matrix of Axelrod's Tournament

From the programming point of view TFT is very simple indeed, in comparison with some other submitted programs. Yet it turned out to be the most successful in the first tournament. Since the success depended on the other programs submitted and their number was fairly small, Axelrod arranged another tournament and received a substantially larger amount of programs: 63 altogether (Axelrod 1980b). This time the participants in the contest were told that TFT would be one of the competing programs. Thus special preparations could be made for confronting TFT. Yet, surprisingly, TFT won again.

Axelrod attributes TFT's success to three factors:

- It is a 'nice' program, that is, it never chooses D first. Thus, against any other 'nice' program it receives 3 in every round.
- It is forgiving, that is, it retaliates a single defection by the opponent by only a single defection. If the opponent after his defection returns to cooperation, then TFT — having once obtained the largest payoff 5 with the opponent getting 0 — returns to cooperation as well.
- TFT did well against certain 'king-maker' programs. These programs did not do very well in the tournament, but caused significant payoff differences between TFT and its main contestants. This makes the success of TFT seem somewhat coincidental.

TFT can be viewed as a pure reciprocity norm: do to your opponent what he has done to you in the previous encounter. Since variants of this norm abound in societies modern and ancient, the question arises as to whether it is possible to explain the emergence and stability of these norms using settings similar to the PD tournaments. In his book Axelrod (1984a) addresses two questions:

1. Under what conditions is TFT a collectively stable strategy?
2. Under what conditions may reciprocity emerge?

The setting investigated is that of a large population of players engaged in mutual interactions modelled as PD games. Each player is characterized by the strategy he resorts to in these PDs. It is assumed to begin with that the entire population adopts the same PD supergame strategy. The first question above, then, pertains to the possibility that an

	C	D
C	R,R	S,T
D	T,S	P,P

Table 3.18: General PD Payoff Matrix

individual member of the population changes his strategy to his benefit. If such a possibility exists, the new strategy that the individual adopts invades the prevailing one. Let us define the concept of invasion.

Definition 3.5 *Strategy A invades strategy B if $V(A \mid B) > V(B \mid B)$, where $V(A \mid B)$ denotes the payoff for the player playing A when the opponent is playing B.*

Collective stability of a strategy adopted by the entire population can now be defined.

Definition 3.6 *A strategy is collectively stable if no strategy can invade it.*

Thus, if each player is doing his best by complying with the prevailing strategy, then the strategy is collectively stable. Stated in yet another way, collective stability of a norm of conduct means that nobody has an incentive to deviate from the norm assuming that the other abide by it. The prevailing situation is, thus, a Nash equilibrium.

Consider a general PD payoff matrix as in Table 3.18. In the table $T > R > P > S$ and $T + S < 2R$. The latter requirement is imposed to prevent the possibility − which might otherwise materialize in experimental settings − of players colluding so that they alternate C and D choices and thus end up with larger cumulated payoffs than by playing C all the time. Axelrod proves the following theorem.

Theorem 3.1 *TIT-FOR-TAT is collectively stable iff*

$$a \geq max \left(\frac{T - R}{T - P}, \frac{T - R}{R - S} \right).$$

We shall not prove the theorem, but just show that the strategy ALL D (that is, unconditional non-cooperation in every game) does not invade TFT if the discount parameter a is large enough. The discount parameter is the weight assigned by a player to the payoff received after one period from the present. If a player receives R in every period of the infinite sequence of PDs, his accumulated payoff is: $R + aR + a^2R + \dots$ For the proof of the entire theorem, the reader is referred to Axelrod's (1984a) book.

Consider the payoff to the ALL D player against the TFT player in an infinitely long sequence of PD games. In the first game he receives T, while the TFT player gets S. Thereafter both players receive P in every round. Thus,

$$V(\text{ALL } D \mid \text{TFT}) = T + aP + a^2 P + a^3 P + \ldots$$

$$= T + aP(1 + a + a^2 + \ldots) = T + \frac{ap}{1-a}.$$

Now the population of TFT players receive the following accumulated payoff:

$$V(\text{TFT} \mid \text{TFT}) = R + aR + a^2 R + \ldots = \frac{R}{1-a}.$$

In other words, ALL D cannot invade TIT-FOR-TAT if

$$T + \frac{aP}{1-a} \leq \frac{R}{1-a}$$

or

$$a \geq \frac{T-R}{T-P}.$$

The theorem answers the first question above, namely that of the conditions for the collective stability of TFT. But what about the emergence of reciprocity in the sense of TFT? Can one also explain why and/or how TFT becomes the prevalent strategy? This is the second question discussed by Axelrod.

Assume that in a population nearly all players choose strategy B. Now, a relatively small group of players choosing A enters the population. Assume, moreover, that proportion p of the interactions of each member of the latter group takes place with another player choosing A. The remaining proportion $1 - p$ of interactions take place with players choosing B. Since players choosing A are relatively rare in the population, nearly all interactions of players choosing B take place with other players choosing B.

Thus, the average payoff to players choosing A is:

$$pV(A \mid A) + (1-p)V(A \mid B).$$

The average payoff to players choosing B is $V(B \mid B)$.

Strategy A invades strategy B iff

$$pV(A \mid A) + (1-p)V(A \mid B) > V(B \mid B).$$

Hence, the invasion succeeds if

$$p > \frac{V(B \mid B) - V(A \mid B)}{V(A \mid A) - V(A \mid B)}.$$

Example. We apply the preceding argument to the supergame consisting of the PDs of Table 3.17 with $B = $ ALL D, $A = $ TFT and $a = 0.9$.

$V(B \mid B) = \frac{P}{1-a} = 10$

$V(A \mid B) = S + \frac{aP}{1-a} = 9$

$V(A \mid A) = \frac{R}{1-a} = 30.$

Substituting these values into the inequality above, we get

$$p > 1/21$$

which means that the invasion of TFT to a population of ALL D players succeeds iff the proportion of interactions between TFT players is larger than $1/21$. This proportion of interactions is sufficient to offset the losses in the first rounds of PDs with the ALL D players. The precise value of p depends, of course, on the particular payoff values of Table 3.17 as well as on the value of the discount parameter.

Axelrod gives a highly stylized reconstruction sketch of how the norm of reciprocity might come about in a normless world. Its main implication for institutional design is in showing how interaction settings may by themselves give rise to norms of behaviour or institutions. Moreover, it shows that in modelling interactions from the viewpoint of norms, the one-shot games may not be adequate since the evolutionary aspects of behaviour cannot be accommodated in that approach.

It is common to interpret TFT as a more 'civilized' or superior norm than ALL D in the PD context. This in itself may be right, but, for example, as a norm of criminal policy, TFT is blatantly in conflict with practices of the modern civilized world. But not only that; it can be shown that TFT's success in PD supergames is based on downright unrealistic assumptions about information transmission. Per Molander shows that when due corrections are made in those assumptions, it turns out that programs superior in performance to TFT exist and, moreover, they are in general more generous or forgiving than TFT (Molander 1985).

Molander's approach stems from the same tradition as Smale's, namely the dynamical systems theory. More specifically, Molander considers the supergame consisting of PDs as a system that moves from one state to another with certain transition probabilities. A state of the system at any given point in time is the pair of choices made by the players. Assume that both players have adopted TFT, but, with a small probability p of error, make erroneous moves. Thus, when TFT requires C to be chosen, the choice actually made is C with probability $1 - p$ and D with probability p. Using the results of the theory of Markov chains, Molander shows that TFT combined with the small error probability leads to the average payoff of $(R + S + T + P)/4$, regardless of how small p actually is. This is a very poor performance in terms of payoffs. In fact, by choosing randomly C and D the players could end up with the same average. Depending on the exact values of the payoffs, ALL D could even do better than TFT when the latter is applied with error probability p. Both Axelrod and Molander suggest that the improvement of TFT's performance in uncertain environments calls for increasing generosity, that is, increasing the probability of the C choice over that dictated by TFT. In other words, sometimes defections of the opponent should be responded to by cooperation. Without going into the details of the argument, we can observe that evolutionary game theory can help us in understanding the emergence of reciprocity norms.

Further reading: Axelrod (1984a); Bartholdi et al. (1986); Molander (1985).

3.12 APPLICATIONS

Some basic game-theoretical tools have now been outlined. In the following we shall illustrate the use of those tools. Since our primary focus is on institutional design, this dictates the nature of the examples. So far we have dealt with two-person games. In the next subsection, however, an institutional solution to an n-person PD will be discussed. Apart from that subsection the n-person game theory will be mainly discussed in the context of constitutions and social choice, both settings that can naturally be modelled as n-person games. The second example illustrates the importance of information asymmetries in finding optimal allocation of resources to various public-sector projects. Since the distribution of information among players is one aspect of institutions that can sometimes be affected by suitable design, the latter example is particularly relevant for our purposes.

3.12.1 Public goods provision game

In view of the fact that PD has been used in modelling collective action problems, such as the emergence of a state, it is natural to question the validity of a two-person game. It is common to respond to this by stating that from the viewpoint of every individual, the situation can be seen as a two-person PD game between that individual and a collective actor — 'the others'. But surely, an n-person analogue of PD could also be envisioned. Its crucial characteristics are:

- The best outcome for each player is one in which he is the sole defector.
- The outcome in which all cooperate Pareto-dominates the outcome in which all defect.
- The worst outcome for each player is one in which he is the only cooperative player.
- Defection dominates cooperation for all players. That is, given any distribution of defectors and cooperators, each player is better off defecting than cooperating.

Consider as an example a particular collective goods provision game described by Russell Hardin (1971). A group of n persons is confronted with the issue of purchasing a divisible collective good, that is, a good that can be purchased in smaller or larger quantities so that when the good is available, an equal amount of it is available to each member of the group. Furthermore, we assume that the good has a constant benefit/cost ratio r. Thus, if one person contributes 1 unit of cost, each member benefits r/n units. For a contributing member the net benefit is $r/n - 1$.

Let now $r = 2, n = 10$, that is, each unit of cost brings 2 units of benefit and there are 10 people in the group. Then we get the payoff matrix shown in Table 3.19.

The payoff matrix differs somewhat from those we have encountered before. Lines marked with (C) and (D) denote the choices available to the individual. The row marked '2' indicates the number of contributors (including the individual whose payoffs are indicated) in case the individual chooses to contribute. The range is from 10 to 1. The row marked '5' gives the possible number of all contributors in case the individual does not contribute. The range is, therefore, from 9 to 0. The uppermost row marked '1' denotes the outcomes $P_i, i = 1, \ldots, 10$ in which the player whose payoffs are indicated in the rows is one of the i contributing players, while $10 - i$ are not contributing. Similarly, the entries $N_j, j = 1, \ldots, 10$ of row marked '6' denote outcomes in which the player is one of j non-contributing players.

The payoffs on line (C) are computed from $(2m/10) - 1$; $m = 10, 9, \ldots, 1$. The payoffs on line (D), in turn, are computed from $2m/10$.

Obviously for all values of m: $2m/10 - 1 < 2m/10$, that is, given any number of other players contributing, C is worse than D as one would expect in an n-person PD. Thus, there is a presumption that no one

1	P_{10}	P_9	P_8	P_7	P_6	P_5	P_4	P_3	P_2	P_1
2	10	9	8	7	6	5	4	3	2	1
(C)	1.0	0.8	0.6	0.4	0.2	0	−0.2	−0.4	−0.6	−0.8
(D)	1.8	1.6	1.4	1.2	1.0	0.8	0.6	0.4	0.2	0
5	9	8	7	6	5	4	3	2	1	0
6	N_1	N_2	N_3	N_4	N_5	N_6	N_7	N_8	N_9	N_{10}

Table 3.19: n-Person PD

contributes and there will be no collective good. This is a variant of the problem focused upon by Mancur Olson in his well-known book (Olson 1965). There it is maintained that the level of provision of collective goods tends in general to be grossly suboptimal.

Hardin suggests that the problem should be approached in the way that many collective goods provision problems are dealt with in practice, namely by collective decision making. More specifically, he builds a model in which the choice of each individual is determined by a majority vote.

The procedure Hardin discusses is the pairwise comparison of all alternative outcomes with the simple majority rule deciding the winner of each comparison. Thus, for example, if N_1 is confronted with P_{10}, there will be nine voters who get 1.0 in the latter outcome, but only 0.8 under the former. So these nine voters presumably vote for P_{10} and N_1 is defeated despite the fact that one voter would prefer it to P_{10}. The vote is taken over all outcomes.

We now construct the preference order of voter h over the outcomes, assuming that whenever P_h or N_h is written, h is assumed to be among the contributors or non-contributors, respectively. From the payoffs it directly follows that the order is the following.

$$N_1$$
$$N_2$$
$$N_3$$
$$N_4$$
$$N_5, P_{10}$$
$$(N_6)P_9$$
$$(N_7)P_8$$
$$(N_8)P_7$$
$$(N_9)P_6$$
$$N_{10}, P_5$$

The security level payoff is 0. This is the minimal payoff the player can guarantee himself unilaterally by choosing 'Don't contribute'. Hence,

P_1, P_2, P_3 and P_4 which are below this level are not feasible. Consequently, their complementary outcomes N_9, N_8, N_7 and N_6 are also infeasible.

The procedure of binary comparisons with majority rule may result in different outcomes depending on the agenda of comparisons, even when the preferences remain the same and each voter votes for his preferred alternative in each comparison. However, in this particular distribution of preferences any agenda produces the same winner. This follows from the fact that there is an outcome that defeats by a simple majority any other outcome. Such an outcome is called the Condorcet winner after the great eighteenth-century French social philosopher the Marquis de Condorcet. A Condorcet winner does not exist in all situations, but in this case P_{10} is the Condorcet winner. This can be seen by going through all the alternatives in the preference list and observing that P_{10} gets a majority in the pairwise contest with any one of them. Thus only 1, 2, 3 and 4 persons out of 10 will vote for N_1, N_2, N_3 and N_4 in pairwise contest with P_{10}, respectively. Similarly, P_{10} defeats P_9 with 9 against 1, P_8 with 8 against 2, P_7 with 7 votes against 3 and P_6 with 6 votes against 4. Moreover P_{10} defeats N_{10} 10 to 0. In the $P_{10} - P_5$ contest the 5 contributors will obviously vote for P_{10}, while the 5 non-contributors are indifferent between P_{10} and N_5. If they abstain, then P_{10} wins by plurality.

Thus, pairwise voting with a simple majority leads here to a Pareto-optimal outcome. Thus it makes sense to resort to collective decision making in deciding about the provision of collective goods. Unfortunately, the pairwise voting with majority rule does not always lead to a Pareto optimal outcome. We shall return to this drawback later on.

Hardin's solution to the n-person PD transforms the game from a non-cooperative to a cooperative one. It is a clear example of an institutional design to solve a behavioural (collective action) problem. The crux of Hardin's procedure is to commit the players to choosing according to the results of voting. The commitment is reasonable since guarantees can be given to voters that the outcomes resulting in lower than security level payoffs will not be subjected to a vote.

Further reading: Hardin (1971; 1982).

3.12.2 Invisible games

The existing institutional framework may be such that although one can quite legitimately speak of a game situation, the game never begins because the player who is to make the first move doesn't do so in anticipation of the response by the other player. Henry Hamburger

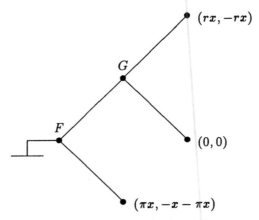

Figure 3.7: White Elephant Game

(1979) uses the concept 'invisible game' to denote such game situations. Some of these games explain the poor performance of some institutions. Otto Keck discusses an example that explains why so many public funds seem to go to waste in new technology developments programmes (Keck 1988).We shall briefly outline Keck's model.

Figure 3.7 describes a two-person extensive form game in which the players are Government (G) and Firm (F). The former has decided to support the development of new technology in the form of grants for research and development. The cost of development is x and Government has announced that it will cover the costs of developing the technology. To encourage the firms, Government also announces that it will give a bonus of πx to the firms that come up with a finished product. Firm, on the other hand, may either accept the offer or inform Government that in its opinion the programme is a waste of funds.

Since it is assumed that Government has already committed itself to support this new technology, it is Firm that makes the first move. It can choose either to accept the grant and get the net payoff πx. If it makes this choice the government's payoff is $-x - \pi x$ since it pays the development costs and the bonus. On the other hand, Firm may decide to blow the whistle, that is tell Government that according to its research or other reliable sources the whole enterprise makes not enough sense to warrant financing. Government may then decide to stop the programme and be finished with it. Both players would then get nothing or lose nothing. Alternatively, Government may decide to pay a reward to Firm, its amount being proportional to x, say, rx. The payoffs are indicated in Figure 3.7 with Firm's payoffs mentioned first.

Now, applying Zermelo's algorithm, we notice that as long as x is a positive number, Government prefers to stop the programme and pay nothing to Firm. Then both get 0 payoff. Knowing this, Firm has no incentive to inform Government that the programme is a white elephant, that is has no such practical value as to warrant its financing. The reason is simple: by accepting the offer Firm gets πx, a sum that certainly exceeds 0. Thus, Firm does not blow the whistle. Government learns about the futility of investing funds for the programme only after the product is finished and in use and then it is too late.

So this institutional framework is not conducive to encouraging feasible, let alone cost-effective, project development. Keck's example is from the West German fast breeder reactor programme in the 1960s, but it is not difficult to envision examples from other fields and contexts. Many aid programmes that are based on good intentions, but have inadequate experience of the cultural, political or infrastructure conditions prevailing in the target area, may lead to huge opportunity costs. Similar arguments apply to science policies of governments. As long as money is coming scientists' way, very few complain about the possible weaknesses of the underlying assumptions concerning the progress of scientific work. Concrete evidence of white elephant games is, however, more difficult to come by because this game is 'invisible'. In other words, it is never played in reality since Firm does not move. Therefore, Government does not necessarily even know that there is a game that could be played.

Invisible games suggest an aspect of institutions that is of crucial importance in their design, namely the distribution of information. In the white elephant game there is something essential that Firm knows, but Government doesn't. One way of modifying the white elephant game into another setting where the support of useless research and development activity would be less likely, is for Government to commit itself to rewarding Firm for revealing the futility of the programme. But the mere cutting out of the other branch emanating from Government's node in Figure 3.7 is not enough, since Firm would still rather accept Government's offer than turn in information about the unfeasibility of the programme, unless Government rewards the latter more than the net utility the programme would bring to Firm. Thus, in order to avoid supporting white elephants, Government has to set $rx > \pi x$. Since white elephants are usually huge projects in financial terms, this solution is unlikely to work. It might even backfire in the sense of encouraging the setting up of all kinds of complicated programme proposals that could then be shot down by firms at considerable profit. It is likely that many kinds of strategic collusions between firms or research institutions would emerge to take advantage of these new funding possibilities.

As a solution to the white elephant problem Keck suggests cost sharing between Firm and Government. In other words, when accepting Government's offer, Firm commits itself to financing part of the work. Since profit can presumably be made from a finished product, Firm has an incentive to participate only in feasible programmes. Thus, with suitable cost and profit sharing the white elephant problem can in some cases be avoided. The exact share proportions depend on the specific characteristics of the problem.

There is a whole spectrum of informational asymmetries that may affect the performance of an institution. For example, the white elephant game is closely related to the principal—agent interaction where the principal does not know all relevant characteristics or information of the person or collective actor whose services he purchases. In some respects the relationship between voters and their representatives is a principal—agent one, thus exhibiting an informational asymmetry.

Another principal—agent setting of asymmetric information is called moral hazard. A typical example of that setting is one in which a person buys insurance for some property that needs occasional maintenance. From the viewpoint of the insurer it would be relevant to know if the buyer gives the property the necessary maintenance and if he exercises adequate care with respect to it. These activities being largely unobservable, the buyer can do something that remains unknown to the insurer. Continuing the insurance example, adverse selection is also one of the principal—agent settings whereby the insurer does not know the relevant properties of the insured. For example, if a person buys health insurance, the insurance company typically wants to know his health condition since this is helpful in risk assessment. To the extent that the insured person knows something about his health risks that is not known to the company, we are dealing with adverse selection.

Further reading: Arrow (1986); Keck (1987; 1988); Rasmusen (1989).

3.13   BIBLIOGRAPHICAL REMARKS

Good introductions to basic game theory from the viewpoint of applications are Brams (1975) and Hamburger (1979). Applications to law are discussed by Baird et al. (1994), to political science by Morrow (1994) and to economics by Rasmusen (1989) and Kreps (1990). Rasmusen's text focuses particularly on the role of information in game theory. Binmore has written extensively on rational decisions and game theory. His *Fun and Games* is a well-written general introduction to basic concepts of game theory using parlour games as examples (Binmore 1992). More advanced comprehensive texts are Myerson's and Fudenberg and Tirole's books (Myerson 1991; Fudenberg and Tirole 1991).

Various equilibrium concepts have been widely discussed over the past decades. Selten's, Myerson's and van Damme's articles are particularly worth studying (Selten 1965; 1975; Myerson 1978; van Damme 1984). Two books have been devoted to the equilibrium selection problematique (Harsanyi and Selten 1988; van Damme 1983).

The ordinal 2 × 2 games are enumerated by Rapoport and Guyer (1966) as well as by Brams (1977). Brams has applied these games to a wide variety of contexts ranging from theology Brams (1980; 1983) to international conflicts (Brams and Mattli 1993). Colomer (1991) makes an interesting application of the ordinal games to the transition of authoritarian regimes into democratic ones. The most recent version of TOM is outlined in Brams's book (Brams 1994).

Inspired by the PD tournaments Axelrod has developed further the ideas concerning the emergence of cooperation and reciprocity (Axelrod 1984a; 1984b; 1981; Axelrod and Dion 1988; Axelrod and Hamilton 1981). See also Leinfellner's essay (1986).

The n-person variants of PD are discussed by Hamburger (1973) and Schelling (1978). The literature of collective action problems modelled as PDs or Chicken games is vast. The reader is referred to Hardin's two books (1982; 1995). The former is a general discussion of collective action problems and how the collectively cooperative choice could be brought about, whereas the latter discusses the sometimes disastrous effects of successful cooperative outcomes, for example in ethnic conflicts. Collective action problems solved at one, say ethnic group, level may cause new ones on a higher, say state, level. The tragedy of the commons problematique is further developed and extended to the problems of managing common pool resources by Elinor Ostrom (1990).

Asymmetric information and the role of reputation in repeated games have been widely discussed in the literature. Selten's chain-store paradox is a widely known contribution to this field (Selten 1978). Wilson (1985) provides a good introduction to the significance of reputation in game models. Interesting political science applications are suggested by Randall Calvert (1987) and by Hannu Salonen and Matti Wiberg (1987).

4 Constitutions

A constitution is, among other things, a set of rules for making collective decisions. As such it is perhaps the best example of an institution aimed at bringing order to interaction between individuals and groups in a society. According to contractarian political theory, the state is the result of a voluntary agreement between individuals who, realizing the collective benefits of an ordered society, surrender some of their freedom in exchange for security against each other. The contractarians do not necessarily maintain that the states have evolved historically through explicit agreements between individuals, but consider the contract as a useful benchmark in evaluating existing institutions. Thus, for example, John Rawls regards arrangements that could have emerged as a result of a voluntary agreement between free and self-interested individuals as just (Rawls 1971).

What, then, constitutes the necessity of social contract or the emergence of the state? Edna Ullman-Margalit (1977) argues that norms come about as responses to social dilemmas which otherwise would become intolerable by causing large collective welfare losses. These dilemmas fall basically into two categories:

- PD situations, and
- coordination problems.

Both dilemmas are solved by setting up appropriate norms. Thus one can expect political constititions to include norms for solving PDs and norms for solving coordination problems.

A coordination problem has two components: the coincidence of interests and ambiguity. The former entails that the parties faced with the problem have no interest in hiding their choices or other actions from each other. On the contrary, the problem is precisely in communicating to the others what one is about to do. Consider the matrix form game of Table 4.1.

	b_1	b_2	b_3
a_1	1,1	0,0	0,0
a_2	0,0	1,1	0,0
a_3	0,0	0,0	1,1

Table 4.1: Coordination Game

Here Row and Column have three options each. There is complete harmony of interests in the sense that each player would like the other to know her choice or, alternatively, know the choice of the other, not in order to use it for her own advantage and to the peril of the other, but in order to benefit both players equally.

Two kinds of equilibria can be distinguished in coordination games. First, a coordination equilibrium which is an outcome such that neither player, or in n-person case none of the players, wishes that any player had made another choice. Second, a proper coordination equilibrium which is an outcome where at least one player would have suffered a loss had any player made a different choice from the one that was actually made. Coordination equilibrium is clearly a Nash equilibrium, while the proper coordination equilibrium is a special case of it.

Now, a coordination problem is encountered whenever there are at least two proper coordination equilibria. A coordination norm is an instruction or rule that prescribes such an action that, if followed by all those involved in the problem, would lead to an equilibrium. Given that others will conform to this rule, each individual player wants to conform to it as well.

PD norms, on the other hand, are of two contradictory types, that is, those that avoid the non-cooperative equilibrium and those that discourage the players from cooperating. The former norms often involve setting up an authority to enforce contracts between parties or sometimes modifying the PD payoff matrix so as to end up with a game that is not a PD any more. The cooperation-discouraging norms, on the other hand, are motivated by the benefits accruing to others from the non-cooperation of the players. Anti-trust legislation or rules encouraging competition between firms are typical PD norms of the latter variety.

Thus game models can explain the emergence of certain types of norms and institutions as solutions to social dilemmas. But supposing that certain activities are to be controlled by the state, what kind of constitution would be rational from the viewpoint of an individual? This basic question is the topic of Buchanan and Tullock's *Calculus of Consent*. We shall briefly outline some of its central arguments in the next section.

4.1   CALCULUS OF CONSENT

The issue of constitutional choice arises in situations where a set of individuals is about to establish a collective body to make decisions that are binding for the individuals. Some crucial aspects of the issue can also be extended to situations involving groups and other collectivities that are about to form higher-order bodies.

Although the general nature of the issues to be decided in the collective body is typically known to the individuals — such issues may deal with trade, security, environment control — the particular items of the agenda are supposed to be unknown. The constitutional choice takes place behind the veil of ignorance, to use Rawls's expression (Rawls 1971). In particular, the individual does not know whether the collectivity will in the future make decisions that are in accordance with her intererests or against them. In the former case, of course, the individual will benefit from the decisions, while in the latter case, she will incur losses.

Buchanan and Tullock consider the constitutional choice problem as one of expected cost minimization(Buchanan and Tullock 1962, 63–84). The individual supports such decision-making rules — that is, constitutions — that minimize the costs she is likely to bear as a member of the body. Two particular sources of loss are considered: the external costs and the decision-making costs. The former are costs that are incurred to an individual as a result of a collective decision that is against her interests. If the decision requires unanimity, this will presumably never happen since the individual can veto such a decision. So, from the viewpoint of external costs, the unanimity rule exposes an individual to zero costs. If, on the other hand, any one member of the collectivity can make collectively binding decisions, then the likelihood of adverse decisions is at a maximum. Let us now define the (collective) decision rule as the number of members of the collectivity that are needed to carry a proposal. Buchanan and Tullock argue that the external costs are normally decreasing with the increase of the decision rule, that is, the larger the decision rule, the smaller the external costs.

The other cost considered by the individual pertains to decision making. To wit, the higher the decision rule, the more individuals have to agree on a stand before it becomes the collective decision . Consequently, the higher the decision rule, the more effort an individual has to put in in order to convince an adequate number of others of the acceptability of her opinion. If she has to convince nobody else, that is, if the decision rule is 1, then there are no decision-making costs, while if she has to get everybody else's approval for her stand, then the costs are at a

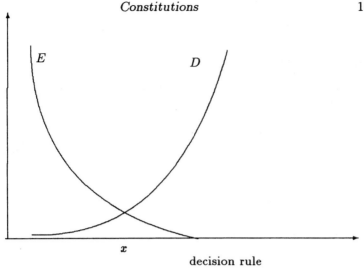

Figure 4.1: External and Decision-Making Cost Functions

maximum. Between these two extremes one would expect the decision-making costs to be increasing with the decision rule. An example of these two cost types as functions of the decision rule is shown in Figure 4.1. Now, the rational choice of a constitution coincides with the value of x for which $E(x) + D(x)$ is at minimum. Here $E(x)$ ($D(x)$, respectively) is the external (decision-making) cost value associated with the decision rule x. Of course, this reflects the favourite position of one individual. Her external and decision-making functions may look like those of Figure 4.1, but the specific details of the functions may well vary from one individual to another. Thus, even though the qualitative properties of the functions may be the same — that is, the external costs may be decreasing and decision-making costs increasing functions of the decision rule — the points at which the sum of those functions is at a minimum may differ for different individuals.

When the functions are reasonably symmetrical the optimal decision rule would seem to be located near $n/2$. This is not necessary, though, but depends on the individuals' subjective assessments of the cost types. Anyway, the simple majority rule $n/2$ rounded upwards if n is odd and $(n/2) + 1$ if n is even seems a very widespread principle. The cost calculus of Buchanan and Tullock gives some support for the rationality of this rule since one could argue that behind the veil of ignorance the individuals by definition do not know the issues to be collectively dealt with and, in particular, do not know whether they are more likely to incur costs in convincing others about the correctness of a stand or in suffering the consequences of adverse collective decisions. This would give some plausibility to postulating symmetric functional forms and, *eo ipso*, to supporting the simple majority rule.

Further reading: Buchanan and Tullock (1962); Taylor (1969); Guttman (1996).

4.2   MAXIMIZING THE CHANCES OF HAVING ONE'S WAY

Stronger cases for the simple majority rule have, however, been built. One of them is the idea that − under certain conditions − the rule maximizes the probability of each individual's being on the winning side. This is Douglas Rae's result and the conditions under which it holds are:

- no individual knows about the future,
- individual votes are mutually independent, that is, an individual's vote does not influence other individuals' votes, and
- each individual is equally likely to support or oppose a proposed policy (Rae 1969).

The first condition follows from the veil of ignorance assumption. The second one is more serious and rarely holds in any real-world voting body. Yet this assumption is often made to simplify probability calculations. On the other hand, it is difficult to see what kind of dependence assumption would be more realistic under the conditions of the veil of ignorance. The third condition seems to follow from the first one.

Taking now a given individual's viewpoint and fixing our attention on a specific issue to be voted upon by the collectivity, we can distinguish four outcomes:

- The proposal which the individual supports is rejected.
- The proposal which the individual opposes is passed.
- The proposal which the individual supports is passed.
- The proposal which the individual opposes is rejected.

Obviously the first and second outcome are unpleasant for the individual and, if anything, she would like to minimize their occurrence. On the other hand, the third and fourth outcomes are the ones in which the individual has her way. To illustrate Rae's argument, consider a situation where there are three individuals, I, II and III, and one issue to be voted upon. The issue is dichotomous, that is, each voter may vote 'yes' or 'no' on the issue. We denote the former vote by 1 and the

	000	(1) Proposal is rejected
	001	(2) Proposal is rejected
Decision rule	010	(3) Proposal is rejected
$k = 2$	011	(4) Proposal is passed
	100	(5) Proposal is rejected
	101	(6) Proposal is passed
	110	(7) Proposal is passed
	111	(8) Proposal is passed

Table 4.2: Vote Vectors in Three-Voter Case

latter by 0. Once the votes have been cast, the voting behaviour can be represented as an ordered triple or vote vector (i, j, k) where $i, j, k = 0$ or 1 and i denotes voter I's, j voter II's and k voter III's vote. Thus, for example, triple $(0, 1, 0)$ means that voters I and III vote 'no', while voter II votes 'yes'.

All possible vote vectors are listed in Table 4.2. Also the outcomes are indicated assuming that the decision rule $k = 2$.

If the agenda of voting is formed by the voters themselves it seems that the outcome (1) is not feasible since no one is going to make a proposal that everybody is opposed to. Thus, there are altogether 7 feasible outcomes. In general with n voters, there are $2^n - 1$ feasible outcomes.

Looking at Table 4.2 from voter I's point of view, two outcomes, namely (4) and (5), are such that she does not get her way. Assuming that the veil of ignorance means that all feasible outcomes are equally likely, this implies that I's probability of not getting her way is $2/7 = 0.29$. With a different decision rule the outcomes are different as well. If $k = 3$, that is, unanimity is required, the proposal is rejected in all but one row of Table 4.2, namely the last one. Voter I does not get her way in 3 cases out of 7. Thus, with probability $3/7 = 0.43$ the voter's interests will not be served. Since $k = 2$ and $k = 3$ are the only larger than majority decision rules for a body with three members, we observe that the simple majority rule minimizes the probability of a voter's not getting her way.

It makes sense to compute only those decision rules that are at least as large as the simple majority, since with smaller rules we would get absurd outcomes in the sense that both those who vote for a proposal and those who oppose it win, that is, a proposal is both passed and not passed. Rae computes the probabilities for values of n ranging from 3 to 12 and it turns out that the simple majority maximizes the probability of an individual's getting her way in all those values of n. In other words, the simple majority rule seems to give the best guarantees of success for individuals under these conditions. This is clearly a strong case for the simple majority rule.

It should be emphasized that the simple majority rule is applied to dichotomous choice situations. The choice is between 'yes' and 'no'. The notorious paradoxes (to be discussed later) emerge when the alternative set is enlarged to three or more alternatives, but the simple majority rule itself operates on two alternatives at a time. We shall now turn to some properties of this widely used rule.

Further reading: Abrams (1980); Rae (1969).

4.3 THE PROPERTIES OF SIMPLE MAJORITY RULE

Consider again a dichotomous voting situation with alternative set $X = \{x, y\}$. We shall denote the vote for alternative x by 1 and the vote for alternative y by -1. Abstaining from voting is denoted by 0. Thus, individual i's vote is a variable B_i that has one and only one of three values: $-1, 0, 1$. A ballot vector is an ordered n-tuple of individual ballots, that is, $B = (B_1, \ldots, B_n)$.

Let now F be a function that can be used in determining the winner as follows:

$$F : B_1 \times B_2 \times \ldots \times B_n \rightarrow \{1, 0, -1\}$$

where $F(B_1, \ldots, B_n) = -1$, 0 or 1 according to whether y or x wins or there is a tie between them, respectively.

F is called a ballot-counting function. Suppose now that F has the following three properties:

1. $B_1 + \ldots + B_n > 0 \Rightarrow F(B_1, \ldots, B_n) = 1$
2. $B_1 + \ldots + B_n = 0 \Rightarrow F(B_1, \ldots, B_n) = 0$
3. $B_1 + \ldots + B_n < 0 \Rightarrow F(B_1, \ldots, B_n) = -1$.

Any such F is then by definition a simple majority rule. The first property says that if those ballots that have been given for x outnumber those that have been given for y, then x is declared the winner. The third property says essentially the same thing, namely if more votes have been given to y than to x, then y is the winner. The second property defines a tie as an outcome where as many votes have been given to x as to y or everyone has abstained. The above three properties define what is meant by the simple majority rule.

Kenneth May proved in the early 1950s that the following characteristics are each necessary and jointly sufficient for the simple majority rule. In other words, the simple majority rule as defined above has all those properties and any rule that has them works exactly like the simple majority rule, that is, makes the same choices. The characteristics are the following:

- Decisiveness: the domain of F is $B_1 \times \ldots \times B_n$.
- Symmetry or anonymity: the permutation of individual ballots leaves the outcome unaffected.
- Duality: if $A_i = -B_i$, for all $i = 1, \ldots, n$, then

$$F(B_1, \ldots, B_n) = -F(A_1, \ldots, A_n).$$

- Strong monotonicity: Let $B_i \geq C_i$, for all $i = 1, \ldots, n$ and $B_i > C_i$, for at least one i. If

$$F(C_1, \ldots, C_n) = 0 \text{ or } 1$$

implies that $F(B_1, \ldots, B_n) = 1$, then F is strongly monotonic.

These are called axioms of the simple majority rule. It is relatively straightforward to see that the simple majority rule as defined above has, indeed, all these characteristics. It is somewhat more complicated to show that any rule that has these characteristics is equivalent to the simple majority rule.

A few comments on May's result are in order. The condition of decisiveness means that F always gives one of the following results: x wins, y wins or there is a tie. This is certainly a useful property since it means that F never fails to give a result. The second property means that just the number of votes counts, not whose votes have been given to each of the alternatives. Duality, in turn, says that if everyone, except those who abstain, changes her mind about the alternatives, then the outcome should also change unless it was a tie before the changes. Strong monotonicity is a requirement that additional votes for the winner should not turn the winner into a loser or produce a tie. Moreover, only one change of mind is needed to break a tie.

These are all *prima facie* plausible properties and May's result thus has by and large a positive ring to it. One may well wonder what drives all those negative results on the simple majority rule that have been given so much attention in the social choice theory. A quick answer to this question is: the number of alternatives. When one leaves the world of dichotomous choice, the simple majority principle turns rather erratic.

Further reading: May (1952); Plott (1976).

4.4   MAJORITY RULE AND STABILITY

In his *Liberalism Against Populism: A Confrontation Between the Theory of Democracy and the Theory of Social Choice* William H. Riker points out that two major obstacles stand in the way of what he calls the populistic view of democracy (Riker 1982):

- the paucity of voting equilibria, and
- the ubiquity of strategic manipulation possibilities.

The former means that the outcomes reached through democratic decision making — that is, voting — are typically not stable. Rather the outcomes — had they been known to the voters beforehand — would have given to at least some of them a reason to vote otherwise than they in fact did. The latter obstacle, in turn, alludes to the possibility of influencing the voting outcomes by either controlling the agenda or by misrepresenting one's opinions in voting. In Riker's view, these two obstacles are both based on the results of axiomatic social choice theory and are thus of the nature of mathematical truths rather than contingent regularities. Therefore, one should look for some other justification of democratic procedures than the intuitive view that they, better than any other procedures, express the will of the people. Riker's own suggestion is the liberalistic view of democracy which amounts to defining democracy as a system in which the people is allowed to change its rulers in elections held at regular intervals.

Further reading: Riker (1982); Nurmi (1984a; 1986; 1987).

4.4.1   Sources of instability

We shall not dwell on the liberalistic view of democracy, but focus instead on the theory that Riker uses in building his argument against the populistic view. In particular, we shall discuss the developments in the study of majority rule that have taken place after the publication of Riker's seminal work. The picture that emerges out of these more recent results is not entirely different from but much more nuanced than the one that Riker could base his argument on. It calls particular attention to qualified majorities rather than simple ones. Hence, it is appropriate to begin with the justifications of the simple majority principle which is often considered to be an operational democratic procedure.

From the days of Condorcet it is known that things get complicated for the majority rule when the number of alternatives is larger than two. There are several reasons for this:

Voter 1	Voter 2	Voter 3
a	c	d
b	b	b
c	d	a
d	a	c

Table 4.3: Preference Profile with a Cycle

1. The Condorcet winner alternative may not be the one that gets most votes, that is, plurality and pairwise majority winners may be different alternatives.
2. The pairwise voting with a simple majority rule may produce a cycle: a beats b, b beats c,..., k beats a.
3. The voters may find it in their interest not to reveal their true preferences in voting. For example, a voter might get a better outcome (from her point of view) if she votes for x rather than y in the pairwise vote between these two alternatives, even though she prefers y to x.

These reasons are illustrated in Table 4.3 which depicts a set of voters and their preferences over a set of alternatives. The latter set is $A = \{a, b, c, d\}$ and the former $N = \{1, 2, 3\}$. The voter preferences are indicated so that the top-most alternative is the most preferred one, the next alternative is the second and so on. Thus, for example, voter 1 ranks the alternatives in the order $abcd$. A description of a set of voters with their preferences over a set of alternatives is called a preference profile.

Upon inspecting the preference profile of Table 4.3, one observes that there exists a Condorcet winner, namely alternative b would defeat all the others in pairwise contests with a simple majority of votes if everyone voted according to their preferences. Yet — and this illustrates problem 1 above — it would not be elected if every voter had just one vote. In fact, it would not get a single vote. So even in a system where a runoff takes place between the two largest vote-getters, b would not be elected since it would not make it to the runoff contest. In this profile it is difficult to say which alternative would win, but presumably it would be one of the set $\{a, c, d\}$.

Regarding the second problem of the majority rule, we observe that should b be eliminated, no difference could be made between the rest of the alternatives, since there is a majority cycle through all three alternatives: a beats c, c beats d and d beats a. In other words, the collective preference relation formed using the majority rule in pairwise comparisons is not transitive but cyclic.

Continuing the illustration using Table 4.3, suppose now that the choice is to be made from the set $\{a, c, d\}$ and that the amendment procedure is being resorted to. According to this procedure, the alternatives are confronted with each other in pairs using a predetermined agenda so that the winner of each comparison goes forward and the loser is eliminated. In this procedure, $k - 1$ comparisons are performed if the number of alternatives is k. Suppose that, according to the agenda, a is first confronted with c, whereupon the majority winner of this contest is confronted with d and the winner of the latter contest is declared the overall winner. This procedure is widely used in contemporary legislatures.

If all voters vote according to their preferences in each contest, the overall winner is d since it defeats a after the latter has beaten c in the first comparison. Anticipation of this outcome — which may be due to information about the preference profile — provides voter 1 whose least preferred alternative is d with an incentive not to reveal her true preference in the first comparison. If instead of voting for a she voted for c, the last contest would take place between c and d, whereupon c would win instead of d. This would be a better outcome for voter 1 and, therefore, her voting behaviour does not necessarily reveal her true preferences.

A crucial role in all these complications is played by the absence of a voting equilibrium, that is, an outcome which exhibits a degree of stability or permanence once it is reached. A paramount example of a voting equilibrium is the Condorcet winner. We shall now turn to literature that discusses the conditions of voting equilibria and the effects of their absence.

Further reading: Riker (1982); Nurmi (1987).

4.4.2   Majority rule in spatial voting games

In spatial voting games the preferences or utilities of the voters are represented as points in a p-dimensional policy space and a distance measure (norm) in the space. Each dimension represents a variable or characteristic of policies that can be adopted. The point representing voter i is the policy that would be ideal from her point of view. Hence it is called her ideal point. We denote it by x_i. Each conceivable policy alternative can also be represented as a point in the policy space. Those points will be denoted by $y_j, j = 1, \ldots$. The utilities of voters are represented by means of the distance measure d in the sense that the larger $d(y_k, x_i)$ is, the smaller is $u_i(y_k)$. Here u_i is voter i's utility function. An example of a spatial configuration of voter ideal points in a two-dimensional

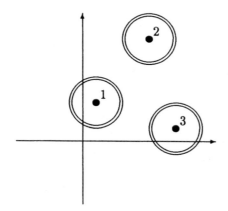

Figure 4.2: Spatial Voting Example

space is given in Figure 4.2. For illustration we could think of a public health insurance policy where the crucial aspects are the percentage of the population covered, represented by the horizontal axis, and the level of benefits, represented by the vertical one. Voter 1's ideal system would have a limited percentage of population covered but the benefits would be fairly high, while voter 2 would consider the ideal policy to have somewhat larger coverage in terms of population and also higher benefit levels. Voter 3, in turn, would ideally cover more of the population than voters 1 and 2, but would prefer low benefit levels.

The voter ideal points are depicted in the figure with dots. Around each dot two circles have been drawn. When the Euclidean norm is used, the distance between two points in the space is measured by drawing a circle with its centre in one of the points and the arc passing through the other. The radius of the circle is then the distance in the sense of this norm. Provided that the voters regard each dimension of equal importance and use the Euclidean norm in measuring distances, the points on each one of the circles of Figure 4.2 represent policies that — if proposed — would be considered equally desirable by the voter whose ideal point is the centre of the circle. Each circle is an indifference curve of the corresponding voter. It represents policies between which she is indifferent.

This setting allows for an investigation of the stability properties of the majority rule. The following three questions can be raised:

1. Under which conditions can one expect to find a majority rule equilibrium outcome?
2. In the absence of an equilibrium, is there a reasonably small subset of alternatives that is likely to contain the outcomes of majority voting?
3. Given an equilibrium outcome, is the majority voting likely to result in it?

There is an extensive literature that addresses the first question, that is, the existence of a majority rule core (see, for example Plott 1967; Davis et al. 1972; Enelow and Hinich 1983; Schofield 1983). This concept is a slightly more general equilibrium concept than that of the Condorcet winner. An alternative belongs to the core iff no other alternative defeats it by a simple majority of votes. Thus the core consists of alternatives that beat or tie all other alternatives.

The results pertaining to the existence of a core imply that with sufficiently high values of p (the number of dimensions of the policy space) the non-existence of a core is generic, that is, 'typical' (Schofield 1983). On the other hand, when the number of dimensions is sufficiently small a core necessarily exists (Greenberg 1979). A straightforward example is the one-dimensional case. The results of Black and Downs demonstrate that under plausible assumptions the median position in a one-dimensional policy space is a voting equilibrium in the sense of the Condorcet winner (Black 1958; Downs 1957).

As to the exact bounds of the number of dimensions that are accompanied by non-existent cores, some results are worth repeating here. Consider first the Plott equilibrium (Plott 1967).

Definition 4.1 *Sufficient conditions for the majority rule equilibrium:*

1. *Indifferent individuals abstain from voting.*
2. *If n is odd (n = number of not indifferent voters), the equilibrium is the ideal point for one voter. If n is even, the equilibrium is the ideal point of an even number of voters or of no voter.*
3. *The remaining voters can be divided into pairs whose interests are diametrically opposed, that is, the members of the pairs would like to move the outcome in precisely opposite directions.*

Assuming that each voter's preferences can be represented by distances from her ideal (or bliss) point measured using the Euclidean norm (the longer the radial distance, the smaller the utility), Plott constructs ideal point configurations for an odd number of voters by fixing one voter's, say voter i's ideal point x_i and then drawing $n/2$ lines through x_i so that the following condition is met: on each line there is an ideal point on each side of x_i (see Plott 1967; Saari 1997). This construction guarantees that x_i is the majority rule core since each line divides the ideal points into two groups so that no more than $n/2$ is on either side of the line. The configuration where a voter's ideal point is the core is in Saari's terminology called the bliss—core point configuration (Saari 1997). Other cores are called non-bliss ones.

Using Plott's construction one can, thus, argue that (bliss—core point) configurations with a nonempty core exist whenever $p \leq (n-1)/2$. However, the slightest perturbation of any ideal point might make the core, thus constructed, disappear. Hence, one cannot expect that cores of this type would be generic in spatial voting games.

The issue of generic nonexistence of cores is addressed by Banks (1995), McKelvey and Schofield (1986) and Schofield (1980). The setting assumed in these studies is more general than that of the n-dimensional Euclidean space in which the voter preferences are representable by Euclidean norms. Now the assumption is that the preferences can be represented by smooth utility functions in the space and that the partial derivatives of all orders of those functions are continuous.

Banks's basic result with regard to the majority rule states that when $p > (n+1)/2$ the majority rule core is generically empty. In other words, if the number of policy dimensions exceeds $(n+1)/2$, there is generally no majority rule equilibrium. Banks's bounds are derived using the singularity approach earlier adopted by Schofield (1980) and McKelvey and Schofield (1986). The bounds are not necessarily the exact ones, since using this approach means losing some information about the cores. Utilizing this lost information more fully gives sharper bounds for the generic non-existence of a core.

With the recent results of Saari, the issue regarding the generic existence of cores in majority voting games in spatial contexts has been resolved (Saari 1997). In fact, Saari's results pertain to a much larger class of voting rules, namely q-rules, but at present we shall just mention their implications for the simple majority rule. In particular, the following are necessary and sufficient conditions for generic existence or nonexistence of a majority rule core:[1]

- For an even number of voters a majority rule core (a bliss—core point or a non-bliss—core point) generically exists if and only if $p \leq 2$. For odd n, the bound is $p \leq 1$.
- Given a fixed value p_* of p, there exists a number of voters n_0 so that for all numbers $n \geq n_0$, the majority rule core is generically nonempty.

---

[1] Technically Saari's results assume strictly convex preferences, but are extendable without essential changes to smooth preferences as well (see Saari and Simon 1977).

The message of these results is clear: one cannot expect a core to exist generically, unless one can fiddle with the number of voters or dimensions. So, given that the non-existence of a core is a typical setting, what can one expect the majority rule to accomplish? Is there a subset of voting outcomes that, given voter preferences, would be likely to contain the results of voting?

Prima facie, it would seem that Pareto-suboptimal outcomes cannot be feasible under majority rule, since by their very definition they are characterized by the property that for any one of them, say y, there exists another outcome, say z, which is not Pareto-suboptimal and which is preferred to y by *all* voters. Surely it would seem that y has no chance of being the voting outcome. Yet the famous 'chaos' theorem of McKelvey states that under certain conditions y may well be the outcome of majority voting (McKelvey 1976; 1979). More to the point, there is nothing in the majority rule *per se* that excludes such a y from becoming the outcome. In fact, according to the theorem, considering an arbitrary point x in the space as the *status quo* and another (possibly the same) arbitrary point y as the desired outcome, one can construct a sequence of points $z_1, \ldots, z_m$ so that $z_1 = x$ and z_j defeats z_{j-1} by a simple majority, (for all $j = 2, \ldots, m$) and $z_m = y$.

This result holds under quite general spatial contexts. In fact, only continuous utility representation of voter preferences is assumed. According to the prevalent interpretation, the result states that under a very general class of circumstances there are no limits to agenda manipulation. Whoever controls the agenda controls the voting outcomes under majority rule. This, however, overstates the practical importance of the result. There are various factors that in practice limit the possibilities of the agenda setter. One of them is that in order to manipulate successfully, she has to know the voter preferences in detail. More importantly, the voters have to be myopic to the point of naivety in their behaviour in pairwise votes.

Despite these reservations, the important theoretical point stands: the majority rule by itself gives no guarantee that the final voting outcomes are anywhere near the voter ideal points. If the simple majority voting outcomes in the real world voting bodies exhibit 'stability', this must be due to special circumstances or constraints. Thus for example, a committee structure might essentially mean a dimension-by-dimension consideration of issues and, thus, even majority voting outcomes would be stable, given the underlying committee structure. This is the idea on which Shepsle and Weingast's structure-induced equilibrium is based (Shepsle and Weingast 1981). Similarly, it could be argued that in many real-world voting situations the apparently multi-dimensional policy space can effectively be regarded as two-dimensional with dimensions representing economic goods so that the set of feasible outcomes is constrained by the amount of available funds or technology. In these types of situations a variation of the median voter theorem can be shown to hold (McCubbins and Schwartz 1985). Thus the stability of outcomes may result from the real-world restrictions on the general assumptions underlying the simple majority core non-existence results. Yet the instability results are undesirable enough to give a good reason to look for other rules that might in general behave in a less arbitrary fashion.

4.4.3 The q-rules

In any collective decision making situation where a fixed rule is being applied, there must be a subset of voters which − if unanimous − determines the collective choice. Otherwise, the body could hardly be called a decision-making one. When the simple majority rule is resorted to, any coalition with at least $(n+1)/2$ members is decisive, that is, determines the collective choice or preference between any two alternatives. When one member, say i, of the collectivity is a dictator, the set of decisive coalitions consists of precisely those coalitions where i is a member. Of course, dictatorial choice rules are not of interest in democratic institutions. The same observation can be made of collegial rules, that is, rules in which there is a small subset of voters that is included in every decisive coalition.

The q-rules are defined by the set of decisive coalitions so that a coalition is decisive iff it consists of at least q members. In other words, any unanimous set of at least q voters can impose its will upon the collectivity and no set of fewer than q members can accomplish the same (Banks 1995; Ferejohn and Grether 1974; Saari 1997; Schofield 1984a; 1984b).

While the results on the existence of majority rule core are generally negative, analogous results on q-rules are basically positive. In fact, by making q very large, one can guarantee the existence of a non-empty core. Clearly, the set of all voters has to be decisive since otherwise something other than the voter opinions would determine the voting outcome. Consequently, setting $q = n$ guarantees a non-empty core which coincides with the Pareto set or the set of Pareto-optimal outcomes. Since the latter always exists, so does the core.

The unanimity rule extensively employed in international organizations has thus a theoretical justification: it is the rule that guarantees the existence of a core. By making choices resorting to this rule one can rest assured that the welfare of no group member can be improved without reducing the welfare of some other member. But, of course, this method of making decisions is extremely conservative: once the Pareto set is reached in the voting process, no further changes are possible. Thus, with $q = (n + 1)/2$ we cannot in general be sure of the existence of the core and the outcomes may be 'chaotic', whereas with $q = n$ the core always exists. What about the intermediate values of q, that is, values $(n + 1)/2 < q < n$? To answer this question, one should first recall the definition of the core: the core consists of those alternatives that cannot be beaten in a pairwise contest by any other alternative that is considered better by at least q voters. Suppose that we replace q in this definition with $q + 1$. In any preference profile, those alternatives that can be beaten by at least $q + 1$ votes can *a fortiori* be beaten by at least q votes as well. Conversely, those alternatives that cannot be beaten by q votes cannot be beaten by $q + 1$ votes either, while there may be some alternatives that cannot be beaten by $q + 1$ votes (and that thus belong to the $q + 1$-rule core), but can be beaten by q votes (and that thus do not belong to the q-rule core). Consequently, the $q + 1$-rule core is always a superset of the q-rule (Saari 1997).

The size of the core can thus be made as small as possible by fiddling with the value of q. If in an effort to avoid extreme conservatism one aims at the minimum number of alternatives in the choice set, one should look for the smallest value of $q \geq (n + 1)/2$ that makes the q-rule core non-empty. Since this can be established by a fairly intuitive argument, one may well ask why this is not done in practice. The answer is straightforward: to compute the optimal value of q one needs to know the preference profile. In general, $q + 1$-rule core is different from the q-rule core.[2] Thus, there is no fixed value of q that would do the trick in a given decision-making body unless the preference profile remains always the same. Moreover, even if the rule could be decided separately for each decision, we still face the basic problem of institutional design, that is, how to make sure that the voters reveal their true preferences if they know that this information is used in the determination of the decision rule. Characterizing what a reasonable solution concept for voting games might look like is one thing and finding a procedure that will always lead to that solution is another. Kramer's result on spatial voting games suggests that, given a preference profile, it is possible to delineate a proper subset of the Pareto set so that under certain behavioural assumptions the voting outcomes will eventually reach this subset and remain within it (Kramer 1977). The behavioural assumption states that the agenda of pairwise voting is built so that each item to be subjected

---

[2] Saari shows that it is generically unlikely that these two cores are the same (Saari 1997).

to pairwise comparison with the previous one (the current *status quo*) has the maximum support. The model underlying this result, thus, is one of electoral competition with the electorate occupying fixed positions in the space and the challenger candidates positioning themselves optimally *vis-à-vis* the incumbents. Kramer's rule that guarantees the convergence of the voting outcomes to a proper subset of the Pareto set is based on maximizing the minimum support of each alternative in the preference profile. For a given set A of alternatives and alternatives x and $y \in A$, let $n(x, y)$ denote the number of voters who prefer x to y. Then

$$v(x) = \min_{y \in A} n(x, y)$$

is the minimum support that x has in all pairwise comparisons in the profile. Define now

$$m* = \max_{x \in A} v(x).$$

Kramer's rule is the q-majority rule with $q = m*$. This rule would then converge to a proper subset of the Pareto set if the agenda were based on vote-maximizing considerations. It is fairly straightforward to see that Kramer's solution set is reduced to the Condorcet winner and thus to the core when one exists.

The majority rule core is the most important solution concept in the theory of voting. It has an obvious game-theoretic background: a core is *eo ipso* a Nash equilibrium. In other words, supposing that all other voters stick to their voting strategies that lead to the core outcome, no voter has a reason to change her mind about her strategy. But in the absence of a core one needs to resort to q-rules with values of q larger than $(n + 1)/2$ to guarantee non-arbitrariness of voting outcomes. In practice, a value of q is fixed and used in all voting situations. We shall discuss one particularly widespread value, namely $q = 2/3$ in the following.

4.4.4 The 2/3 and other qualified majority rules

The generic existence of a core for q-rules, provided that the value of q is large enough, gives a theoretical justification for greater than simple majority rules. The practical problem is, however, that of deciding which particular value of q should be chosen. In some parliamentary contexts the value $q = 2(n + 1)/3$ has been used. This was also the rule adopted in 1179 by the Catholic Church in the election of the pope (see Saari 1995). Although this rule seems completely *ad hoc*, it has been given the plausible justification that in order for a candidate to replace the current incumbent the latter has to lose the support of at least half of his previous

voters	issue 1	issue 2	issue 3
voter 1	yes	yes	no
voter 2	no	no	no
voter 3	no	yes	yes
voter 4	yes	no	yes
voter 5	yes	no	yes

Table 4.4: Anscombe's Paradox

supporters (Saari 1995). Should the value of q be smaller, say just a little higher than the simple majority, only a few former supporters of the incumbent could upset the balance. Thus the position of the incumbent would be more precarious if the rule were near the simple majority, 2/3 being the minimum value of q that guarantees that at least half of the support of the incumbent has to vanish for a new incumbent to replace him. Or, stated in terms of the majority required for legislative proposals, at least half of the support of the existing legislation has to be dissatisfied with it *vis-à-vis* the new proposal.

The rule of 3/4 can also be given a justification, as follows. Wagner (1983) shows that this rule avoids Anscombe's paradox (Anscombe 1976). The paradox itself can be described with the aid of Table 4.4 (Gorman 1978).

Here we have five voters giving their votes on three dichotomous issues. If the majority rule is used, 'yes' wins on issues 1 and 3, while 'no' wins on issue 2. The paradox consists of the observation that a majority of voters (voters $1 - 3$) is on the losing side on a majority of issues.

A crucial role in the paradox is played by the simple majority rule since it determines the winning stand on each issue. Wagner's theorem states that if, instead of the simple majority rule, one requires a 3/4 majority for a proposal to be adopted, then it cannot be the case that a majority of voters is on the losing side on a majority of adopted proposals. If the adoption of a proposal requires a 'yes' vote of at least 3/4 of the voters, then in Table 4.4, 'no' would win on all issues. Thus, more than a majority of voters would be on the losing side. However, the crux of the theorem is that the majority of issues must be computed in the subset of issues in which the 'yes' stand wins or the proposal is adopted. This subset is, of course, empty in this example. Thus the example is not well chosen since 'yes' would win by a majority on issues 1 and 3, whereupon only one voter, namely voter 2 is in the minority on a majority on issues. More to the point is the example of Table 4.5. In Table 4.5 voters 1, 2, 5 and 7, that is, a majority, are in the minority

voters	issue 1	issue 2	issue 3
voter 1	yes	no	no
voter 2	no	no	yes
voter 3	no	yes	yes
voter 4	yes	yes	yes
voter 5	yes	no	no
voter 6	yes	yes	yes
voter 7	no	yes	no

Table 4.5: Modified Anscombe's Paradox

on two adopted proposals out of three if the simple majority is being applied. With the rule of 3/4 the subset of adopted proposals is still empty.

In another article Wagner shows that the rule of 3/4 is in fact an instance of a more general principle which states that if the number of 'yes' entries in an $n \times k$ matrix (n voters, k issues) is at least $(1 - \alpha\beta)nk$, then no more than βn voters are in the minority on more than αk adopted issues (Wagner 1984). The rule of 3/4 thus turns out to be a special instance of this more general rule with $\alpha = \beta = 1/2$.

The results on q-rules suggest that qualified majorities are effective ways of ensuring stability of the outcomes in situations involving a large number of issues. On the other hand, extremely high majority thresholds make all changes in the status quo unlikely. In a way, then, each system has to sail between the Scylla of extreme conservatism and the Charybdis of simple majority 'chaos'.

The negative results on majority rule have often been interpreted to suggest that the only important thing for democracy is that whoever is in charge can be ousted in periodic elections. Even assuming this liberal view of democracy — which is by no means uncontested — does not, however, provide sufficient grounds for simple majority rule in parliamentary decision making. In higher-dimensional policy spaces majority rule equilibria are not typical and, hence, in all likelihood the outcomes reached are unstable and more or less accidental. In addition to higher majority thresholds, multi-chamber systems have been suggested as ways of enhancing the stability of voting outcomes (Riker 1992). Indeed, in many contexts higher than simple majority thresholds and bicameralism are alternative ways of achieving stability in outcomes. Without taking a stand on the issue of which one of them is in general more efficient, it can be argued on the basis of the preceding results that a case can be built for super-majoritarian legislative coalitions.

For institutional design the main message of the results discussed, however, is that the peculiarities of the simple majority rule provide no grounds for arguing that democratic ways of decision making are random or chaotic in general. Rather the recent work gives a much more nuanced picture. If stability and intuitive consistency are sought, these are possible to achieve through the use of q-rules.

The discussion on voting procedures has thus far focused on pairwise comparisons of alternatives using either simple or qualified majorities. There are, however, other methods of making social choices some of which utilize more fully the information contained in the individual preference relations by taking into account the positions that various alternatives occupy in the preference orders. We turn next to some of these methods. Since the variety of the procedures is vast, we shall limit ourselves to a brief discussion of the best-known methods and discuss some general results that are pertinent in the design of voting and electoral institutions.

4.5   BIBLIOGRAPHICAL REMARKS

As Jon Elster observes in his article, there is, surprisingly, no literature that tries to explain in positive (that is, non-normative) terms the constitution-making process (Elster 1996). His own contribution therefore deserves to be mentioned as a useful guide to literature that might be helpful to an understanding of the process of constitutional design. Specific constitutional topics, such as the emergence of decision rules, conventions and property rights have, on the other hand, been dealt with in the literature (Coleman 1990; Sugden 1986).

Buchanan and Tullock's classic book is still a useful introduction to the subject of constitutional design (Buchanan and Tullock 1962). So is Buchanan's monograph which dicusses the problematique of individual freedom and the powers of the state from a libertarian perspective (Buchanan 1975). In a somewhat similar spirit Robert Nozick outlines a defence of a state that enforces contracts and protects individual property, but abstains from redistribution (Nozick 1974). The veil of ignorance concept, which often comes in handy when one is claiming that an arrangement or institutions is just, was originally employed by Rawls in his discussion about the principles of social justice (Rawls 1971).

Constitutional arrangements typically involve the choice of an electoral system along with the organization of the supreme decision-making bodies. These issues are discussed in Giovanni Sartori's book (Sartori 1994). The interaction of party and electoral systems is discussed in two monographs (Katz 1980; Shugart and Carey 1992). There is an extensive literature on the advantages and disadvantages of various electoral systems. A modern classic is Rae's book (Rae 1967) which raised a number of topics followed up in a volume edited by Grofman and Lijphart (1986). A comprehensive description of the existing electoral systems is given by Dieter Nohlen (1978). Rein Taagepera and Matthew Shugart have written a very illuminating book on the properties of various electoral systems (Taagepera and Shugart 1989). Also the volume edited by Lijphart and Grofman (1984) gives a good overview of the variety of systems and their properties.

The majority rule is axiomatized in May's article (1952). Also Plott's (1976) overview article discusses this rule. The chaos theorems are proven and discussed by Richard McKelvey (1976; 1979). He and Norman Schofield derive some core existence results (Schofield 1984a; 1984b; McKelvey and Schofield 1986 which are improved upon by Jeffrey Banks (1995). Donald Saari's article gives a good overview of the previous results and provides some important new ones (Saari 1997).

5 Social Choice

We begin with voting paradoxes. Although *prima facie* unrelated to institutional design, they are worth discussing since they exhibit problems that have been encountered in devising and evaluating voting institutions. Moreover, the paradoxes have provided the motivation for much of the present theory of institutions. It could even be argued that the first impetus to institutional design came from attempts to avoid certain types of anomalies in the outcomes reached. This is definitely the case with regard to the first paradoxes to be dealt with in the following.

5.1 VOTING PARADOXES

The theory of social choice has a very long — albeit discontinuous — history. The first observations about the properties of voting procedures can be traced back to antiquity, but serious systematic work began in the late eighteenth century during the years preceding the French revolution of 1789.

5.1.1 The Borda paradox

Historically the first of the two paradoxes discovered in this period was one that nowadays is called the Borda paradox. It was presented to the French Royal Academy of Sciences by Jean-Charles de Borda. Faithful to the spirit of the Enlightenment, Borda was not content with pointing out a serious problem in the behaviour of an existing institution — namely the one-man-one-vote or plurality system — but proposed a solution to the problem. The latter is called the Borda count today. Let us first focus on Borda's example of the paradox that haunts the plurality system (see Table 5.1).

Suppose that each person votes according to his true preferences, that

Voters 1−4	Voters 5−7	Voters 8,9
A	B	C
B	C	B
C	A	A

Table 5.1: The Borda Paradox

is, sincerely. This assumption was energetically defended by Borda who implied that it would not be honourable to act otherwise. With sincere voting and the plurality principle A wins since it gets 4 votes, while B gets 3 and C 2 votes. However − and this is the crux of the paradox − in pairwise comparisons A is defeated by both B (with 5 votes to 4 votes) and C (5 to 4 as well).

Borda uncovered two defects in the plurality rule:

1. that it may elect a 'wrong' candidate in the sense that the chosen candidate would be beaten by all the others in a pairwise contest with a majority of votes, and
2. that it may fail to elect a 'right' candidate, that is, one that would beat all the others with a majority of votes (Condorcet winner).

Borda was mainly preoccupied with the first problem, that is, avoiding that the 'bad' candidate is chosen. In his view A is not the right choice because it would be defeated by a majority in pairwise comparison with all other alternatives. Such an alternative is called the Condorcet loser. That Borda was mainly interested in the first problem is evident from the fact that his proposal for the voting procedure does not necessarily avoid the second problem. In other words, it is possible that a Condorcet winner is not elected when the Borda count is used. This flaw was pointed out to Borda by his contemporary the Marquis de Condorcet.

The Borda count utilizes the entire preference profile in determining the winner. If the number of alternatives is k and if each voter has a complete and transitive preference order with no ties, the alternative ranked first by a voter gets $k-1$ points from him, the next one $k-2$ and so on. The sum of points that the voters have given to an alternative is its Borda score. The alternative with the largest Borda score is declared the winner (Borda winner). So in Table 5.1 the Borda winner is B which is also the Condorcet winner.

In this example, then, the Borda count seems to elect the Condorcet winner. This is, however, not always the case (an example can be found in Table 5.7). On the other hand, the Condorcet loser is never elected

Voters 1,2	Voters 3,4	Voters 5,6	Voter 7
D	A	B	D
C	D	A	C
B	C	D	B
A	B	C	A

Table 5.2: Some Peculiarities of the Borda Count

by the Borda count. Thus the first problem of the plurality voting is solved by the system.

On closer inspection the Borda count seems to have some rather undesirable properties. Table 5.2 gives examples.

The Borda scores are D 15, A 10, B 9 and C 8. Consequently, the collective preference order formed on the basis of the scores is $DABC$. Suppose now that D is not eligible. We now recompute the Borda scores by simply ignoring D. The new scores are A 6, B 7 and C 8. Thus the collective preference order is CBA. It turns out that the order between A, B and C is reversed when D is removed.

It may also happen that the removal of the alternative with the lowest Borda score reverses the order among the remaining ones. Fishburn shows that the Borda winner is not necessarily the Borda winner in any but one of those proper subsets of alternatives of which it is a member (Fishburn 1974). Suppose that x is the Borda winner in the set X. Fishburn's result says that x has to be a winner also in one proper subset of X, but not necessarily in more than one such subset. In other words, the Borda winner may be rather 'unstable'.

The Borda count can be seen as an attempt to solve a problem related to the plurality voting system. Borda's system is, however, only a partial solution. It fails on what is called the Condorcet winner criterion, that is, the criterion that requires that the Condorcet winner be chosen, whenever one exists. It took about one hundred years before a system was invented that would solve both problems of plurality voting. In the early 1880s E.J. Nanson devised a scheme that is based on sequential application of the Borda count (Nanson 1883). The specific aim of the system is to choose the Condorcet winner when such an alternative exists. This is accomplished by computing at each stage of a sequence of rounds the Borda scores of all alternatives that have not yet been eliminated. In each round the alternatives with Borda scores less than equal to the average score are eliminated. That the end result is necessarily the Condorcet winner — if one exists — is due to the fact that the Borda score of a Condorcet winner is necessarily higher than the average. Thus it belongs to the set of non-eliminated alternatives

I	II	III
A	C	B
B	A	C
C	B	A

Table 5.3: The Condorcet Paradox

at each stage. Nanson's method thus solves both problems of plurality voting. Borda's and Nanson's methods are surely the earliest successful attempts at systematic design of voting systems.

Further reading: Fishburn (1974); Colman and Pountney (1978); DeGrazia (1953); McLean (1996); Nanson (1883).

5.1.2 The Condorcet paradox

The paradox that carries Condorcet's name is much better known than the Borda paradox.[1] An example of it is shown in Table 5.3. The essence of the paradox is that the collective preference relation over the alternatives formed on the basis of majority comparisons may be cyclic. Consider the following agenda of balloting:

First ballot: A vs B
Second ballot: the winner of the first ballot vs C.

According to the usual amendment procedure, the winner of the second ballot is declared the winner. Assuming that the voters vote according to their preferences in each ballot, C is the overall winner. However, suppose that the third ballot were arranged between C and B, that is, between the winner of the second ballot and the loser of the first one. Then, B would win. Thus the majority comparisons result in a cycle: A beats B, C beats A and B beats C. Stated in another way, the paradox consists of the observation that no matter which alternative is picked from the set $\{A, B, C\}$, a majority of voters would prefer some other alternative to the chosen one.

In the same way that Borda proposed a method to improve the performance of voting institutions, Condorcet too suggested a way to overcome the problem of cyclic majorities. His method has probably never been used in practical decision-making contexts. This, incidentally, is true of Nanson's method as well. Condorcet's suggestion is to analyse

---

[1] Condorcet's paradox is sometimes called the paradox of voting. This title is, however, somewhat undeserved since there are other equally serious paradoxes related to voting. To avoid confusion we shall use the more appropriate label 'the Condorcet paradox'.

the sizes of the majorities forming a cycle and to disconnect the cycle at that particular link which is the weakest. The intuition underlying this stratagem is that the observed majorities reflect imperfectly 'the true will of the people'. If the latter is assumed to be complete and transitive, it makes sense to look for the smallest possible modifications in the observed pairwise comparison voting results that would make the ensuing majority preference relation transitive. Surely, this provides no solution to the version of the Condorcet paradox presented in Table 5.3 since each link between alternatives is equally strong, but this is not always the case.

H. Peyton Young argues that Condorcet's suggestion is identical to the method that is known as Kemeny's method (Young 1988). Given a preference profile the method aims at constructing a complete and transitive preference relation that would be closest to the observed individual preferences. This is done by first listing all possible complete and transitive preference relations and then counting the number of pairwise preference changes that would be needed in order to make each of the possible preference relations unanimously acceptable. In other words, for each constructed complete and transitive preference relation, the following questions is posed: how many individuals have to change their minds about preference between some pairs of alternatives to have their modified preferences coincide with the constructed one?

To illustrate Kemeny's method, consider again the Borda paradox (see Table 5.1 on page 125). For the computation of the Kemeny winner we list all possible preference orders that are complete and transitive. Thereafter we determine the required preference changes (Riker 1982, 80):

1. ABC: this coincides with the preferences of the first 4-voter group. The 3-voter groups needs to change their preferences between A and B as well as between A and C to agree with ABC. This calls for 6 changes. In the 2-voter group all three preferences have to be reversed. Thus, altogether 12 preference changes are needed to make ABC a unanimously accepted ordering.
2. ACB: the left-most group needs 4 changes, the middle group 9 changes and the right-most one 4 changes. Altogether 17 changes are needed.
3. BAC: $4 + 3 + 4 = 11$ changes.
4. BCA: $8 + 2 = 10$ changes.
5. CAB: $8 + 6 + 2 = 16$ changes.
6. CBA: $12 + 3 = 15$ changes.

The preference order that calls for the smallest number of preference changes in order to become unanimously acceptable is thus BCA. This

group	issue 1	issue 2	issue 3	party supported
A (20%)	X	X	Y	X
B (20%)	X	Y	X	X
C (20%)	Y	X	X	X
D (40%)	Y	Y	Y	Y

Table 5.4: Ostrogorski's Paradox

is the Kemeny and Condorcet solution.

Further reading: Abrams (1980); Black (1958); McLean and Urken (1995); Riker (1982); Young (1988).

5.1.3   Ostrogorski's paradox

The preceding two classic voting paradoxes are excellent examples of institutional design problems and attempts at their solution. We shall now move on to other paradoxes related to voting institutions for which the solutions are either not as yet known or are of a rather different nature from those touched upon above. As it turns out they are all problems related to direct versus representational forms of democracy. They can also be viewed as cross-level inference paradoxes.

At the beginning of this century Moise Ostrogorski (1903) discovered a paradox which nowadays bears his name. One of its variations is presented in Table 5.4. The table describes a hypothetical contest between two parties, X and Y, in a constituency consisting of four groups of voters. The groups A, B and C comprise 20 per cent of the voters each, while 40 per cent of the electorate belong to group D. In issues 1 − 3 the groups are able to identify the party which is closer to their opinion. Supposing that these three issues are of crucial and equal importance in determining each voter's party choice, it is plausible to assume that the voters belonging to groups A, B and C vote for X since X better than Y represent their opinions on a majority of issues. Similarly the voters belonging to group D would support party Y since its stand is closer to theirs on all three issues. Thus the election is won by X since it gets the support of 60 per cent of the voters. Notice, however, that if each issue were decided separately, Y would be victorious in every issue, getting 60 per cent of the votes.

We shall call the procedure whereby one first amalgamates over rows and then over columns of the table the issue−person or IP amalgamation, in contradistinction to the procedure in which one first determines the majority winning party on each issue and thereupon the winner by

voters	office 1	office 2	office 3
1−3	D	D	D
4	D	D	R
5	D	R	D
6	D	R	R
7−9	R	D	R
10−12	R	R	D
13	R	R	R

Table 5.5: Paradox of Multiple Elections

determining that party which is victorious on a majority of issues. The latter procedure is called person−issue or PI amalgamation. Now, one interpretation of Ostrogorski's paradox is that it occurs whenever IP and PI amalgamations result in different outcomes. This is certainly not the only possible interpretation, but certainly a plausible one.

It is difficult to see how Ostrogorski's paradox could be solved in an analogous way to the preceding paradoxes, that is, by suggesting a specific method which would avoid the problems crystallized in the paradoxes. Rather the crux of the paradox is in the observation that representational systems may end up with predominantly contradictory policies to those that would ensue from issue-by-issue direct policy choices of the electorate.

Further reading: Bezembinder and Van Acker (1985); Daudt and Rae (1978); Lagerspetz (1995); Ostrogorski (1903); Rae and Daudt (1976).

5.1.4   The paradox of multiple elections

A compound majority paradox of more recent origin is the paradox of divided government discussed by of Brams *et al.* (1993; 1997). Suppose that 13 voters can choose between a Democratic (D) and Republican (R) candidate for (1) the House of Representatives (office 1), (2) the Senate (office 2) and (3) the governor (office 3) (see Table 5.5). Obviously, R is elected to 1, D to 2, and D to 3. However, none of the voters chooses strategy: R for office 1, D for office 2, D for office 3 (RDD).

It is easy to see that Table 5.5 involves Ostrogorski's paradox as well: IP amalgamation results in R, while PI amalgamation ends up with D. Not all Ostrogorski's paradoxes involve the paradox of divided government, however. A case in point is Table 5.4 where (YYY) is the strategy chosen by the largest number of voters. In fact, the paradox of divided government can be viewed as an extreme version of Anscombe's paradox (see Table 4.4 on page 120). The latter can be reformulated as follows: there exists such a majority of issues that the set of voters who are on the winning side on all those issues is smaller than a half of the whole voter set. As a special instance, it may happen that the majority of issues consists of all issues. Furthermore, it may be the case that the set of voters who are on the winning side on all issues is empty. When this is the case, we have an instance of the paradox of divided government. Thus the paradox of divided government is a particularly strong version of Anscombe's paradox.

The source of Anscombe's and Ostrogorski's paradoxes can, as Bezembinder and Van Acker have pointed out, be traced to the non-associativity and non-bisymmetry of the majority rule. Consider a fixed profile of individual preferences over a set of alternatives. Let $M(x, y)$ denote the majority value of x and y, that is, the element of the pair that is supported by the majority.

Associativity of relation M would require that

$$(M(M(x, y), z) = M(x, M(y, z))$$

for all x, y and z. This requirement does not, however, hold for the majority rule. An example of the non-associativity of M is Condorcet's paradox.

Bisymmetry, on the other hand, requires that

$$M(M(x, y)), M(w, z)) = M(M(x, w), M(y, z))$$

for all x, y, w and z. Clearly, the majority rule does not satisfy this requirement, either.

Some work has been done in finding institutions that could handle the paradox of multiple elections (Brams et al. 1997). The issues voted upon in referenda may well be interrelated so that the outcome of one issue determines to some extent the desirability of the alternatives on other issues. In such a case an optimal way to proceed is to conduct the referenda separately on each issue so that the voters know the results of the preceding referenda before participating in the next one. This procedure would often be impractical. A somewhat more realistic suggestion is one that, in a referendum consisting of k issues, calls for the voters to indicate which ballot k-tuples they would accept. This would amount to applying the approval voting method to multiple-issue referenda. The winners would then be determined as those ballot k-tuples which have received the largest support. Various other social choice methods could also be applied.

Further reading: Bezembinder and Van Acker (1985); Brams et al. (1997).

Opinions	MP's 1—8	MP's 9—12	Vote total
'yes'	5	11	84
'no'	6	0	48

Table 5.6: Referendum Paradox

5.1.5   The referendum paradox

All representative democracies in which the referendum institution is at least occasionally resorted to are potentially vulnerable to the paradox which consists of a clear majority of voters voting for an alternative (for example 'yes') and a clear majority of the representatives voting for its negation ('no'). Of course, there is nothing paradoxical in this if the representatives know something that the voters do not know (or, for that matter, if the voters know something that the representatives do not know). But it may well happen even if the representatives consider themselves just agents of the majority of their supporters and thus do not claim to have access to any classified or otherwise special information on the matter in question. Consider the following example (Table 5.6).

The electorate consisting of 132 persons sends one representative for each eleven persons to the twelve-member parliament. Each representative considers himself an agent of the majority of his constituency. Thus, 8 MPs are in favour of the 'no' alternative, while 4 MPs support the 'yes' alternative. Yet, in the electorate as a whole, 'yes' has a handsome 64 per cent majority.

This paradox undermines a relatively common institution, namely the consultative referendum. The basic rationale of this system is that the voters are consulted in a matter that will finally be decided by another authority, typically the parliament. If now the MPs regard themselves as representatives of the majority of their supporters, the referendum paradox creates a clear norm conflict among the legislators. Thus it seems that this paradox can be considered as a case against the consultative referendum, in contradistinction to the binding one.

Further reading: Nurmi (1997a; 1997b).

5.1.6   Effect of the preference aggregation method

The choice of voting method is often considered to be of minor importance. Yet in certain situations the outcomes may crucially depend on the method that is being used. Table 5.7 gives an — admittedly extreme — example of how the method determines the choice in some preference profiles.

4 voters	3 voters	2 voters
a	b	c
c	c	d
d	e	e
c	d	b
b	a	a

Table 5.7: 5 Alternatives, 5 Winners

Let us consider five relatively well-known procedures, four of which are widely used. Let us, moreover, see which alternative(s) these procedures would choose in the preference profile of Table 5.7 assuming that each voter votes according to his true preferences.

1. Plurality voting or one-person-one-vote system. This results in a since the largest number of voters rank a first.
2. The plurality runoff system: if one alternative gets more that 50 per cent of the votes, then it is elected. Otherwise, a second round is held where there are two alternatives, namely the two largest vote-getters in the first election. The candidate with the largest number of votes is elected on the second round. In this example, the plurality runoff winner is b, since no alternative exceeds the 50 per cent threshold on the first round. Thus the second round is needed with a and b running. In the second round b is elected as the right-most group prefers b to a.
3. The Borda count: the winner is e.
4. The Condorcet winner: although this is not a voting system as such we notice that there is a Condorcet winner, namely c, in this profile. Thus, any system that necessarily chooses a Condorcet winner (for example the amendment procedure) chooses c.
5. The approval voting system is one in which each voter may give each alternative either 1 or 0 votes and the alternative with the largest vote sum is the winner. Supposing that in this profile each voter of the left-most voter group gives 1 vote for each of the three highest ranked alternatives, while all other voters give 1 vote for their two highest-ranked alternatives, the winner is d.

Thus all five systems result in a different winner in the example of Table 5.7. In the case of approval voting an *ad hoc* assumption was made concerning how many alternatives the voters would vote for. However, the point of the example is to show that voting procedures may produce different outcomes and, thus, it is worthwhile to consider their systematic properties. Two ways of proceeding in this work can be discerned in the literature. The first branch of research focuses on specific voting systems, determines their properties or sets up criteria of good performance, and finds out whether the systems satisfy these criteria. The other more influential tradition operates on a more general level and deals with theoretical properties and their mutual compatibility or incompatibility. We shall discuss both of these branches of inquiry, beginning with the more specific one.

Further reading: Nurmi (1988b).

5.2   THEORETICAL COMPARISON OF SINGLE-WINNER VOTING SYSTEMS

Voting procedures can be evaluated on several criteria. We have already touched upon two of them, namely the Condorcet winner and the Condorcet loser criterion. The former is the requirement that the procedure should invariably result in the Condorcet winner when there is one in the preference profile under consideration. The latter, in turn, imposes the restriction on the choice resulting from the procedure that it should not contain the Condorcet loser if such an alternative exists according to the profile. Both are at the first sight quite reasonable requirements.

Table 5.8 gives a summary comparison between eleven voting procedures using seven criteria. This is a fairly small subset of those used in the literature, but gives an idea of the relative advantages and disadvantages of various procedures. The evaluations reported in the table assume that the profile of true preferences of the voters is given, whereupon the winners determined by various procedures can be found out. These winners are then compared with the various restrictions imposed by the criteria. In Table 5.8 '1' means that the procedures represented by the row is compatible with the criterion represented by the column and '0' means that the procedure may at least sometimes lead to a conflict with the criterion. The Condorcet winner criterion is denoted by a and the Condorcet loser criterion by b. Thus, for example, the procedure called Copeland is compatible with both a and b, while plurality is compatible with neither of these two criteria.

Before proceeding further in the list of criteria let us briefly introduce the procedures. The amendment procedure, plurality voting, the Borda count, approval voting, plurality runoff and Nanson's method have already been outlined in the preceding. Copeland's procedure is based on pairwise comparisons of alternatives using the simple majority rule.[2]

---

[2]Copeland's procedure can be applied in any context where a round-robin tour-

Voting system	Criteria						
	a	b	c	d	e	f	g
Amendment	1	1	1	1	0	0	0
Copeland	1	1	1	1	1	0	0
Dodgson	1	0	1	0	1	0	0
Maximin	1	0	1	1	1	0	0
Plurality	0	0	1	1	1	1	0
Borda	0	1	0	1	1	1	0
Approval	0	0	0	1	0	1	1
Black	1	1	1	1	1	0	0
Plurality runoff	0	1	1	0	1	0	0
Nanson	1	1	1	0	1	0	0
Hare	0	1	1	0	1	0	0

Table 5.8: A Comparison of Voting Procedures

The alternative that defeats the largest number of others is the Copeland winner.

Dodgson's method is an extension of the Condorcet winner concept. When determining the winner of Dodgson's procedure one looks for the minimum number of individual preference changes that are needed to make any given alternative the Condorcet winner. That alternative which needs the smallest number of such changes is declared the winner. The Dodgson winner is an attempt to look for an alternative that is as close as possible to a Condorcet winner.

The maximin procedure is also based on pairwise comparisons of alternatives. For each alternative one determines its minimum support when it is confronted with the $k - 1$ other alternatives. Once these minimum votes have been found, that alternative which has the largest minimum support is declared the winner. Hence the label 'maximin', that is, one maximizes the minimal support.

Table 5.8 indicates that these four procedures are all consistent with the Condorcet winner criterion. This is not surprising since they are all based on pairwise comparisons and, thus, utilize a setting in which the eventual Condorcet winner is likely to outperform the others. They are, however, not all able to exclude the Condorcet loser. The demonstration of incompatibility of a procedure with a criterion typically amounts to producing a preference profile in which the restriction imposed on the

---

nament can be arranged. What one needs are data on which one of any pair of alternatives defeats the other unless there is a tie. Copeland's system or some variation of it is actually being used in many sport tournaments.

1 voter	1 voter	1 voter
a	*c*	*b*
d	*d*	*c*
b	*b*	*d*
c	*a*	*a*

Table 5.9: Maximin Method Results in the Condorcet Loser

choices by the criterion is not satisfied by the outcome of the procedure. As an example of such a demonstration consider Table 5.9. (For other demonstrations, see Fishburn 1977; Richelson 1979; Straffin 1980; Nurmi 1987). In this example all alternatives, including the Condorcet loser *a*, have the same minimum support, namely 1. Thus, they are all elected. Consequently, the method does not necessarily exclude a Condorcet loser.

The plurality voting, plurality runoff, approval voting and Borda count all fail on the Condorcet winner criterion. We have already given an example demonstrating this failure (see Table 5.7 on page 133). There all four procedures result in choices that do not include the Condorcet winner.

Black's method chooses the Condorcet winner when one exists, otherwise it chooses the Borda winner. It is, thus, a mixture of two criteria. By definition it is compatible with the Condorcet winner criterion. It also necessarily excludes the Condorcet loser since neither the Condorcet winner nor the Borda winner is a Condorcet loser.

Hare's procedure is identical to the well-known single transferable vote system when it is applied to a single-member constituency. Thus, given a preference profile one first determines if there is an alternative that is ranked first by more than 50 per cent of the voters. If there is one, then that alternative is elected. Otherwise, one eliminates the alternative that has been ranked first by the smallest number of voters. This alternative is simply removed from the preference profile so that the second-ranked alternative of those voters who ranked the eliminated one first is now considered their first ranked one. One then looks at the 'truncated' preference profile in an attempt to find if there is an alternative ranked first by more than 50 per cent of the voters. If there is, then that one is elected. Otherwise the elimination is continued until eventually a winner is found.

Hare's procedure is not compatible with the Condorcet winner criterion, while it necessarily excludes the Condorcet loser. The latter property follows from the fact that the winning alternative has to be

1 voter	2 voters	2 voters
c	a	b
b	c	c
a	b	a

Table 5.10: Hare's Method Does Not Always Elect the Condorcet Winner

the Condorcet winner in the reduced set of alternatives that is left after the final round of eliminations. Since the Condorcet loser does not defeat any other alternative in a pairwise contest, it cannot be elected. Table 5.10 demonstrates the incompatibility of Hare's system and the Condorcet winner criterion. There c, the Condorcet winner, is eliminated, whereupon b wins.

The third criterion — labelled c — in Table 5.8 on page 135 is the majority winning one. It requires that if an alternative is ranked first by at least 50 per cent of the voters, then it should be elected. This criterion is obviously satisfied by all systems that are compatible with the Condorcet winner criterion since an alternative that is ranked first by more than half of the electorate is *eo ipso* the Condorcet winner. But also some of the procedures that do not satisfy the Condorcet winner criterion — plurality, plurality runoff and Hare's system — are compatible with the majority voting requirement.

From the viewpoint of democratic theory the criterion d is quite crucial. It is called monotonicity. It requires that when an alternative, say x, wins an election, it should also win when its support is increased, *ceteris paribus*, that is, when its position is improved with respect to other alternatives *and no other changes are made in the preference profile*. Fortunately, only two widely used systems — plurality runoff and Hare's — violate this condition.

Criterion e is a Pareto restriction on choices. It says that if all voters strictly prefer alternative x to alternative y, then y is not chosen. It does not insist that x is chosen since there may be alternatives that are better than both x and y. This criterion is also satisfied by most systems, the exceptions being the amendment procedure and approval voting.

On the other hand, consistency is a property that relatively few procedures possess. In elections resulting always in a single winner, it imposes the following requirement. Suppose that once the ballots have been cast, they are put into two baskets, N_1 and N_2. The results are computed separately for ballots in these baskets. Suppose that x wins in N_1 and in N_2. Then, consistency requires that x should also win when the ballots in N_1 and N_2 are computed together. Although this requirement seems at first sight pretty obvious, it is satisfied only by plurality and approval voting as well as by the Borda count. For all other systems of Table 5.8 profiles can be found where the consistency is violated.

The final column g in the Table 5.8 denotes the criterion that has been been called property α or heritage in the literature. It requires that if an alternative wins in a large set of alternatives, then it should also win in all proper subsets of the set. Of the systems in Table 5.8 it is satisfied by only one, namely approval voting, and even then under the additional assumption that the voters have dichotomous preferences (accepted and not accepted alternatives) that do not change in the subsets.

Although Table 5.8 presents a relatively small subset of criteria used in comparing voting systems, it shows that in designing such procedures one has to accept tradeoffs and think carefully about the relative importance of criteria. Table 5.8 is also limited with regard to the procedures investigated. Another more common approach to the design of social choice institutions focuses on the properties of systems and their mutual compatibility. We shall now outline a few well-known results from this extensive literature.

Further reading: Brams and Fishburn (1983); Fishburn (1977); Nurmi (1987); Richelson (1979); Straffin (1980).

5.3   SOME SOCIAL CHOICE THEOREMS

The social choice theory is well-known for its numerous incompatibility results. Typically these state that procedures possessing certain intuitively desirable properties do not exist, that is, the achievement of one nice property excludes the achievement of another. The information contained in the results is very important in the design of voting systems or resource allocation mechanisms. In the following we shall briefly outline some incompatibility results which are widely referred to in the literature. To get an idea of how one goes about proving the results, we also sketch the proofs of some of them.

Since the results focus on somewhat different formalizations of an institution, we shall begin with the basic vocabulary. Three distinct models of an institution are discussed in the social choice literature:

1. Social welfare function. This is a function that maps n-tuples of individual complete and transitive preference relations into a set of complete and transitive preference relations.
2. Social choice function. This maps the $n + 1$-tuples consisting of individual preference orderings and subsets of alternatives into the set of subsets of alternatives.

3. Social decision function. This is a singleton-valued social choice function. Sometimes the social decision functions are called resolute social choice functions.

Unfortunately, the terminology is sometimes a little confusing. So, for example, in some works the social choice function is defined as a social decision function, while the term 'social choice correspondence' is reserved for social choice functions in the above sense. The precise definitions and sometimes the context reveal which particular content each term has.

5.3.1 Arrow's theorem

Arrow's theorem is by far the best-known result in the social choice theory. It deals with social welfare functions. Accordingly, it assumes that an n-tuple of individual weak preference relations — all complete and transitive — is given and one wants to construct an institution that would aggregate these into a complete and transitive collective preference relation so that the end result adequately reflects the individual preferences. What more specifically is meant by 'adequate reflection' is captured by the following conditions:

1. unrestricted domain,
2. independence of irrelevant alternatives,
3. Pareto condition,
4. non-dictatorship.

The first condition states that the function should be defined for all n-tuples of individual preferences. In other words, the social welfare function should give a result, no matter what kind of preference profile is encountered. The second condition requires that for any pair x, y of alternatives, whether x is collectively weakly preferred to y or *vice versa* depends on the individual preferences between x and y only. Thus the relationship that x and y have with some third alternative z should have no bearing on the social preference between x and y. The third condition has, in fact, been introduced in the preceding, but in somewhat different terms. In this context the Pareto condition states that if every individual strictly prefers an alternative to another, then the former should also be regarded in the collective preference relation as at least as preferable as the latter. The usual construal of the non-dictatorship condition has it that there should not be such an individual whose preference dictates the social preference between all pairs of alternatives regardless of the preferences of other individuals. Arrow's theorem can now be stated.

Theorem 5.1 *No social welfare function satisfies conditions 1–4 and results in a complete and transitive collective preference order.*

Sen's proof of the theorem proceeds via two lemmas and a couple of definitions (Sen 1970). First the concept of almost decisiveness of an individual with regard to a pair of alternatives is defined.

Definition 5.1 *An individual i is almost decisive for x against y iff $x \succ_i y$ and $y \succ_j x$ for all other individuals j imply that $x \succ y$, where '$\succ$' denotes the collective preference relation.*

In other words, an individual is almost decisive for one alternative against another one, iff his preference dictates the social preference for this pair of alternatives if all the other voters have a preference that is opposed to his preference. Thus, being almost decisive for one alternative against another means that if the person in question is unanimously opposed by all others, his preference prevails concerning that pair of alternatives. However, should any other individual share his view concerning the preference of that pair of alternatives, any collective preference is possible.

Being decisive for one alternative against another is a much more informative property.

Definition 5.2 *An individual is decisive for an alternative against another one iff his preference over the pair prevails, no matter which preferences the other individuals have.*

The first lemma used in the proof of Arrow's theorem says that should any individual be almost decisive over any pair of alternatives and should all the conditions of the theorem hold, then this individual is decisive over all pairs of alternatives. The second lemma then shows that, in any preference profile, there is an individual who is almost decisive over some alternative pair or else some of the conditions of the theorem are violated.

The theorem has inspired a vast literature. The conditions that it shows to be incompatible have been scrutinized one by one for plausibility. If any one of them is removed, then it is not difficult to find procedures satisfying the rest. Thus, for the theorem to have any bite at all, the conditions need to be fairly desirable. Yet the plausibility of several conditions can be questioned.

With regard to the universal domain condition, one could argue that all preference profiles cannot possibly be solved in a reasonable way anyway, so why couldn't we accept some mild domain restrictions and thus get away from the incompatibility? Another objection could be directed against the independence of irrelevant alternatives. In some cases it might seem downright unjust not to take into account the preference intensities. Thus one could argue that there are situations in which one knowingly gives up the second condition.

Even the non-dictatorship condition has been criticized, not because one would accept dictators, but because the definition of a dictator differs from the intuitive meaning of the term. This has recently been pointed out by Keith Dowding (1997). He points out that the Arrovian dictator is not an individual who determines the outcomes, but one whose opinion happens to coincide with the outcomes reached. The Arrovian dictator is not necessarily even aware of being one. Thus one could even argue that the absence of an Arrovian dictator would lead to a very strange situation where nobody's preference coincides with the social preference.

Further reading: Arrow (1963); Dowding (1997); Kelly (1988); Sen (1970).

5.3.2   The Gibbard–Satterthwaite theorem

One of the indicators of cumulativeness of a field of research is the occurrence of simultaneous or at any rate independent discoveries of significant facts. One such fact − or rather a theorem − was independently established by Alan Gibbard (1973) and Mark Satterthwaite (1975). The theorem deals with social decision functions in contradistinction from social welfare functions. It pertains to manipulability of these functions, that is, to the question whether one can in general expect the individuals to reveal their true preferences in voting or resource allocation (bidding, selling, buying, and so on).

For notational convenience we shall here replace '$\succeq$'− the symbol of weak preference − with 'R'. The subscript indicates the individual whose preference relation is at issue. Let us define (individual) manipulability of a social decision function with regard to an alternative set X and preference profile $R = (R_1, \ldots, R_n)$ as follows.

Definition 5.3 *A social decision function F is manipulable in situation* (X, R) *by individual i iff i strictly prefers choice set*

$$F(X, R_1, \ldots, R_{i-1}, R_i^m, R_{i+1}, \ldots, R_n)$$

to choice set

$$F(X, R_1, \ldots, R_{i-1}, R_i, R_{i+1}, \ldots, R_n),$$

where R_i^m *is different from i's true preference* R_i.

It is noteworthy that the only difference between the two profiles of the definition is in i's preference: R_i is the true one, R_i^m the 'manipulated' one. The manipulability of a social decision function simply requires that an individual and a situation can be envisioned in which the social decision function is manipulable by that individual.

The Gibbard–Satterthwaite result is negative in suggesting that all reasonable procedures may contain incentives for the individuals not to reveal their true preferences or other opinions. The content of 'reasonable' will be spelled out in the following definition.

Definition 5.4 *A social decision function is non-trivial (non-degenerate) iff, for each alternative x, a preference profile can be found such that x will be chosen.*

What is thus excluded is any such function that would never result in a given alternative, no matter what preferences the individuals express. This certainly seems a minimum requirement for collective decision making. The Gibbard–Satterthwaite theorem is the following.

Theorem 5.2 *Every universal and non-trivial social decision function is either manipulable or dictatorial.*

The proof of the theorem as presented by Alan Feldman proceeds by focusing on the nearly simplest possible collective choice situation, namely one involving two voters and three alternatives (Feldman 1980). It is first shown that any universal, non-trivial and non-manipulable social choice function must satisfy the Pareto condition in a situation involving only two persons. Thereafter, one goes through all 36 different preference profiles – both voters have six different preference rankings over three alternatives – and determines the possible winners excluding outcomes that do not satisfy the Pareto condition. It turns out that the ensuing outcomes are either manipulable at some profiles or the choice function makes one of the voters a dictator.

The theorem is perhaps less dramatic than it seems. What it says is that in all systems one may encounter situations in which at least one individual would benefit from not revealing his true preference, provided that the others do. In other words, the strategy of revealing one's true preferences does not invariably lead to a Nash equilibrium. The informational requirements of a successful preference misrepresentation – that is, giving 'false' information about one's preferences in order to get them better served by the ensuing outcomes – are very stringent. One has to know quite a bit about other voters' preferences and voting strategies to be able to manipulate the outcome.

Intuitively, voting systems differ with regard to the difficulty of manipulating the outcomes. In some systems all one basically needs to know is the distribution of the first-ranked alternatives of other voters.

In other systems much more detailed information about the preference profile would typically be required. There are also differences with regard to the amount of computational resources one needs for manipulation. Neither of these observations undermines the Gibbard—Satterthwaite theorem, but they have important implications for institutional design. Some literature dealing with the informational and computational requirements of manipulation in various systems already exists (Bartholdi et al. 1989; Bartholdi and Orlin 1991; Chamberlin 1985; Nurmi 1990).

   Further reading: Bartholdi et al. (1989); Feldman (1980); Gibbard (1973); Satterthwaite (1975).

5.3.3   Gärdenfors's theorem

The Gibbard-Satterthwaite theorem deals with social decision functions or resolute social choice functions. Since many common voting systems allow for ties, the result is not immediately applicable to those systems. However, it can be shown that all systems of Table 5.8 on page 135 are manipulable (Nurmi 1984b). Moreover, a fairly dramatic incompatibility theorem pertaining to social choice functions has been proven by Peter Gärdenfors. It deals with anonymous and neutral systems, that is, systems in which no individual or alternative is discriminated for or against.

Theorem 5.3 *Let F be a social choice function defined for at least three voters. If F is anonymous, neutral and satisfies the Condorcet winner criterion, then F is manipulable.*

The proof proceeds by starting with a particular three-voter, three-alternative profile where one alternative is ranked first by two voters and postulating that this alternative ought to be chosen from that profile. Another profile is then focused upon. In particular, all logically possible choice sets from the latter profile are analysed. For each such choice set one then shows that if this were the actual choice, then the procedure would be manipulable in some other profile (Gärdenfors 1976). Since in the first profile there is a Condorcet winner, it is by definition elected by any procedure that satisfies the Condorcet winner criterion. Thus all systems satisfying this criterion are also manipulable.

   A crucial role in this theorem is played by profiles where two alternatives are ranked equally high by a voter. In fact, should a more stringent requirement be imposed on voter preferences, namely that no ties are allowed, we can find procedures that avoid the incompatibility. In other words, with strict individual preferences one can find procedures that satisfy anonymity, neutrality and the Condorcet winner criterion and are still non-manipulable.

Example 1. Let X be a set of alternatives, P a profile of strict preferences and F a social choice function, so that
$F(X, P) = \{x\}$ if x is the Condorcet winner,
$F(X, P) = X$, otherwise.
This clearly satisfies the Condorcet winner criterion and can also be shown to be non-manipulable (Gärdenfors 1976). Its main flaw is that it gives no guidance as to the choice when a Condorcet winner does not exist. The following is a slight improvement.

Example 2.
$F(X, P) = x$ if x is the Condorcet winner,
$F(X, P) = B$, otherwise,
where B is the set of Pareto-undominated alternatives.

The incompatibility of the Condorcet winner criterion and other desirable criteria has been established by many other authors besides Gärdenfors. We shall single out two important examples from these results.
Further reading: Gärdenfors (1976).

5.3.4   Young's and Moulin's theorems

A glance at Table 5.8 (page 135) reveals that consistency is a relatively uncommon property among the voting procedures discussed. Indeed, it seems to characterize only those systems that do not resort to pairwise comparisons in determining the winners. More specifically all those systems that are compatible with the Condorcet winner criterion seem to fail on consistency. On the basis of just a limited sample of procedures, a general claim that one of these properties would exclude the other cannot be made. However, Young proves that one cannot have a procedure that both always elects the Condorcet winner when one exists, and is consistent.

We have already dealt with the concept of consistency in single-winner systems. Let us now give a more general definition of it.

Definition 5.5 *Let X be the alternative set. A social choice function F is consistent iff for all partitionings N_1, N_2 of the voter set N, the fact that*

$$F(X, R^1) \cap F(X, R^2)$$

is non-empty, implies that

$$F(X, R^1) \cap F(X, R^2) = F(X, R),$$

where R^1, R^2 and R refer to N_1's, N_2's and N's preference profile, respectively.

Consistency thus requires that no matter how we divide the voters into groups to compute winners in those groups, if there are common winners of those groups, they and only they ought to be winners in the entire voter set as well.

Young's theorem is the following.

Theorem 5.4 *All anonymous, neutral and consistent social choice functions violate the Condorcet winner criterion.*

In other words, what one observes in looking at the Condorcet winner criterion and consistency columns in Table 5.8 is not accidental, but a necessary connection. Of course the connection would not matter very much if consistency were not a desirable property. But it is since it enables decentralized mechanisms to be used in preference aggregation. Thus consistent systems are particularly suitable for decentralized institutions.

Hervé Moulin's theorem deals with a particularly important issue of democratic theory, namely the incentive to vote. In a very entertaining article Fishburn and Brams (1983) discuss the possibility that a group of voters could by abstaining end up with a better election result than by voting according to their preferences. Since abstaining can be considered as a particular form of preference misrepresentation, the result makes a finer distinction within the large class of systems that are known to be manipulable. Let us consider an example of what is known as the no-show paradox (Table 5.11). Suppose that the plurality runoff system — or, what in three alternative cases amounts to precisely the same thing, Hare's system — is being used. With sincere voting, that is, everyone voting according to his true preferences, A wins (since no alternative gets more than 50 per cent of votes in the first round, there will be a second round with A and B running, whereupon A wins). Suppose now that 25 out of the 40 voters with the preference order BCA had not voted at all. In that case, none of the alternatives would have received more than 50 per cent of the 75 votes and the runoff would have taken place between A and C. In that runoff C would have won. Clearly C is better than A from the viewpoint of the abstainers. Thus the 25 would have been better off by not voting than by voting.

Since one of the goals of designing voting systems is to give voters a reason to vote, the systems where the no-show paradox may occur are rather suspicious. Unfortunately, Moulin's theorem tells us that if vulnerability to the no-show paradox is our main preoccupation, we cannot have systems that satisfy the Condorcet winner criterion. More specifically the theorem says the following.

Theorem 5.5 *All procedures that satisfy the Condorcet winner criterion*

35 voters	25 voters	15 voters	25 voters
A	B	B	C
B	C	C	A
C	A	A	B

Table 5.11: The No-Show Paradox

are vulnerable to the no-show paradox.

In other words if one wants to avoid the no-show paradox, one has to settle for systems that do not necessarily elect Condorcet winners. The theorem says nothing about systems that do not satisfy the Condorcet winner criterion. In particular, it says nothing about the plurality runoff or Hare's system. Table 5.11 shows that these systems are, indeed, vulnerable to the no-show paradox. Yet Table 5.8 reveals that they do not satisfy the Condorcet winner criterion. Their showing is thus poor on both counts.

Further reading: Moulin (1988); Young (1975).

5.4   BIBLIOGRAPHICAL REMARKS

A good way to start studying social choice is Black's seminal work which not only gives an introduction to the problems dealt with in the theory but also contains essential historical material (Black 1958). Especially useful are extracts from Condorcet's and C.L. Dodgson's (*alias* Lewis Carroll's) works. A more comprehensive treatment of the history of social choice theory is provided by McLean and Urken (1995). Of course, Arrow's path-breaking work as well as Fishburn's and Sen's monographs are well worth studying (Arrow 1963; Fishburn 1973; Sen 1970). Kelly's book contains a large number of results in a similar *genre* as Arrow's (Kelly 1978). MacKay, in contrast, devotes a whole book to interpreting Arrow's theorem (MacKay 1980). Kelly (1991) has also published a very extensive bibliography of social choice theory. Thomas Schwartz's book is relatively demanding, but contains much useful material not found elsewhere (Schwartz 1986). The same applies to B. Mirkin's book (Mirkin 1974).

The various paradoxes are dealt with more extensively in the works cited in the preceding. The applications of the social choice theory to political science are presented for example by Shepsle (1974), Straffin (1980), Riker (1982) and Nurmi (1987). Shepsle's early work deals with various reactions to Arrow's theorem. Straffin outlines a set of criteria for evaluating voting systems. Riker's book is devoted to issues stemming from democratic theory. Plott's overview article is also useful (Plott 1976). Donald Saari's relatively recent book is a very well-written text for anyone interested in the mathematical intricacies of the theory of voting (Saari 1995).

Social choice theory is currently a very extensive field with several subfields. Three of the latter are particularly worth mentioning although they are only briefly touched upon in this book: the spatial models, models based on individual choice functions and probabilistic models. The classic treatise in the first subfield is Anthony Downs's book (1957), while more recent developments are reported by Enelow and Hinich (1984; 1990). The best variant choice problem, which is easily translated into a social choice one, has been the focus of research originated by M.A. Aizerman. The book by Aizerman and Aleskerov (1995) gives a good overview of the results. The probabilistic social choice theory is outlined, for example, by Intriligator (1973) and Coughlin (1992).

6 Strategic Voting and Institutional Design

The founding fathers of the theory of voting, the Marquis de Condorcet and Jean-Charles de Borda, were primarily interested in improving specific voting procedures. They took their point of departure in what they thought was the obvious rationale of voting, namely to find out the will of the voters. Although Condorcet was keenly aware of the possibility that a vote might not reveal the voter's true will, he did not consider the act of voting as strategic. Rather he and, even more explicitly, Borda viewed what we nowadays call strategic misrepresentation of preferences as something dishonest and not worthy of serious study. This view largely prevailed until the 1960s (Nanson 1883; Black 1958). A notable early exception is C.L. Dodgson, who discussed strategic voting in the 1870s (Farquharson 1969).

The advent of game theory and the rather widely held suspicion, reinforced by the negative results of simple majority rule, that collectivistic notions like the will of the people are devoid of meaning, cast doubt on the assumption of sincere voting. If no such thing as the will of the people exists, why not assume that the voters try to further their own altruistic or egoistic interests in voting? In his pioneering work Farquharson (1969) approached voting as a game of strategy (see also Dummett 1984; McKelvey and Niemi 1978; Moulin 1983). In this view each voting system and situation opens a set of voting strategies for every voter. Given voter opinions or preferences over the decision alternatives, some voting strategies are more plausible than others. The criterion of plausibility on which Farquharson focused his attention is admissibility or dominance. Thus, he wanted to find out whether the voting outcomes would be determinate or uniquely predictable assuming that the voters act strategically in the sense of resorting to only admissible (undominated) voting strategies.

Farquharson's path-breaking work was followed by a stream of results — discussed above — pointing out that not only did the assumption of the voters being strategic instead of sincere make a difference *vis-à-vis* the voting outcomes, but that it was a possibility which could not be remedied by switching to new and different voting systems. Thus Gibbard and Satterthwaite showed that all non-discriminatory and non-trivial voting systems that can be viewed as (singleton-valued) social choice functions are vulnerable to strategic behaviour (Gibbard 1973; Satterthwaite 1975). In a similar vein, Gärdenfors proved that an unpleasant incompatibility prevails between, on the one hand, the requirement that a social choice correspondence always result in a Condorcet winner when one exists, and, on the other hand, the requirement that the procedure be invulnerable to strategic voting (Gärdenfors 1976).

More recently the study of strategic voting has received a major impetus from the results in mechanism design theory. In this theory the problem of implementing social choice correspondences assumes the following formulation: given a social welfare criterion, design a procedure that results in desired states — desirability being defined by the criterion — in equilibrium. In other words, in mechanism design one explicitly resorts to the game-theoretic notion of equilibrium and imposes the restriction that the desired social states not only be attainable by means of the procedure, but that they are to be attainable as game-theoretic equilibria. This literature provides a wider setting in which the peculiarities of voting procedures can be understood.

We shall next discuss a few examples stressing the intuitive importance of strategic considerations. Then some basic definitions and notation are introduced. Thereafter, some recent results on mechanism design are outlined.

6.1 STRATEGIC VOTING CAN MAKE A DIFFERENCE

Undoubtedly the earliest known discussion about strategic voting can be found in the letter of Pliny the Younger to Titus Aristo at the beginning of the second century AD. The case reported in this letter is used as the primary example by Farquharson (1969) (see also Riker 1982, 173–174; Pliny's letters can be found in Pliny 1969). Pliny was presiding in the Roman Senate when the decision was to be taken as to whether the freedmen of consul Afranius Dexter should be convicted to death (C), banished (B) or acquitted (A). Afranius had been found murdered in his home. His slaves had already been executed, as was the custom in

Group I	Group II	Group III
A	C	B
B	B	A
C	A	C

Table 6.1: First Profile of the Roman Senate

similar cases. There was some uncertainty as to the cause of death of the consul. In particular it was not known if the freedmen — had they in fact been involved at all — had merely assisted Afranius in suicide, that is, simply done their duty. Pliny's most preferred solution was to acquit the freedmen. As the presiding officer in the Senate he was in the position to choose the voting procedure. After carefully pondering upon the possible outcome if the more common pairwise voting method were adopted, he proposed plurality voting. Against Pliny's hopes and expectations, banishment (B) won. Pliny thought that if the pairwise method had been used, the first vote would have been between A, on the one hand, and B and C, on the other, that is, between 'not guilty' and 'guilty'. On this vote, Pliny estimated, the latter alternative would have won. Hence the final vote would have been between B and C, both worse alternatives than A in Pliny's opinion. That Pliny's stratagem of resorting to a less common voting procedure backfired shows either that his prediction concerning the voting behaviour of the other senators was in error or that he was not the only strategic actor in the Senate.

Pliny saw his problem as one of agenda design. With the benefit of hindsight it is, however, arguable that he could have achieved the same outcome by simply resorting to sincere voting. Consider three groups of voters with the following preference profile of Table 6.1.

Let us assume that the groups are of such size that any two of them constitute a majority of senators. Pliny estimated that his group (I) would be slightly larger than the others, but certainly smaller than 50 per cent of the senators. Suppose that instead of forcing the plurality voting system to be adopted, Pliny had resorted to the normal procedure of confronting A first with the rest of alternatives, and, in case A received less than 50 per cent of the votes, then confronting B with C to determine the winning alternative. Pliny surely thought that there would have been a second vote between B and C. Now in this vote, his second-best alternative B would probably have won since both Group I and Group III regarded B as better than C. Although there may be some uncertainty as to the preference relation between A and C in Group III, it is clear that B would have prevailed in the final contest, provided that Group II did not constitute more than 50 per cent of the senators. But, of

Group I	Group II	Group III
A	C	B
C	B	A
B	A	C

Table 6.2: Second Profile of the Roman Senate

course, Pliny could only guess the sizes of each group and, thus, for him it was quite plausible to propose plurality voting since it seemed the only possibility of obtaining the best possible outcome A.

Let us now consider a slight modification of the above example, namely one in which Group I's preference relation between B and C is reversed. Thus we get the preference profile of Table 6.2. This is the Condorcet paradox profile. Suppose that the amendment procedure is used and that the agenda is the following: (1) A vs B, and (2) the winner of (1) vs C. With sincere voting B first beats A, whereupon C defeats B, thus becoming the winner.

This outcome is certainly less than satisfactory for voters in Group III. By resorting to a non-sincere voting strategy so that in the first ballot they vote for A instead of B, they can change the outcome from C to A. The voters in Group I would have no objection to this outcome, whereas the voters in Group II certainly would, but the latter voters have no way to avoid the outcome A.

Consider Table 6.1 again and the same agenda as in Table 6.2. If all voters are sincere, then obviously B wins both the first and the second ballot. Could Groups I and II do something about it (as Group III clearly has no incentive to change the outcome)? Group II could, by voting for A, change the outcome of the first ballot, whereupon A instead of B would confront C in the second ballot. But the second ballot would now result in A, which is a worse outcome than B in Group II's opinion. Thus this group has no incentive to deviate from its sincere strategy. Group I, in turn, has no way of changing the outcome of the first ballot since it was in the minority to start with. Thus outcome B is invulnerable to a similar manoeuvring which upsets the sincere outcome C in Table 6.2. The essential difference between the two tables is that in the former there exists a Condorcet winner. In Table 6.2 no such alternative exists. The presence of the Condorcet winner explains the necessity of its becoming the outcome of sincere voting *per definitionem*, but the fact that it is also the outcome of strategic voting requires a closer look at the theory of voting.

6.2   STRATEGY-PROOFNESS OF VOTING PROCEDURES

6.2.1   Notation and some definitions

The following notation as well as the definitions are standard (Ishikawa and Nakamura 1979; Moulin 1983; Peleg 1984). We shall consider the set N of n voters and a set X of alternatives (candidates, policies, and so on). Each voter $i \in N$ is endowed with a complete and transitive preference relation R_i over X. The set of all possible transitive and complete preference relations of voter i is W_i. An n-tuple of individual preference relations, that is, the preference profile, is denoted by R^N. A social decision function (SDF, for brevity) is a function F from the set of preference profiles to X. The set of all preference profiles involving N is denoted by W^N. A social choice correspondence (SCC), in turn, is a function from W^N to 2^X, the set of subsets of alternatives. Thus SDF always gives a single alternative (winner), while SCCs may produce ties between several alternatives.

Let F be an SDF and $R^N \in W^N$. Then the matrix form game associated with F and R^N is denoted by $g(f, R^N)$ where the strategy set of each $i \in N$ is W_i, f is the outcome function and R_i is i's preference relation over the outcome set X. In other words, the strategies of voters are their reported preferences over the decision alternatives.

An equilibrium point (e.p., for short) of $g(f, R^N)$ is a profile $T^N \in W^N$ if for all $i \in N$: $f(T^N) R_i f(T^{N-\{i\}}, R'_i)$, for all $R'_i \in W_i$. Here $T^{N-\{i\}}$ denotes the profile of $n-1$ preference relations which is obtained from T^N by deleting i's preference relation. Equilibrium point is a Nash equilibrium, that is, in an e.p. no voter regrets having voted in the way she did, provided that the others stick to their strategies.

A strong equilibrium point (s.e.p.), in turn, is a $T^N \in W^N$ if for any coalition S of s voters and for any s-tuple of strategies L^S of the members of S, there exists a voter $j \in S$ who regards $f(T^N)$ at least as good an outcome as $f(T^{N-S}, L^S)$.

If for any $R^N \in W^N$, the profile R^N is an e.p. in $g(f, R^N)$, then F is individually non-manipulable or strategy-proof. In individually non-manipulable systems, it never benefits an individual not to report his true preference relation, provided that the others do. SDF F is called coalitionally strategy-proof if for any $R^N \in W^N$ the strategy R^N is an s.e.p. If an SDF is coalitionally strategy-proof, it is also individually strategy-proof. What is less obvious is that the converse is also true (Ishikawa and Nakamura 1979, Lemma 5.3).

Given the set N of players, we recall the definition of a simple game: $G = (N, V)$ where V is the set of winning coalitions. The core $C(G, X, R^N)$ of simple game G can be defined with the help of the binary dominance relation $D(R^N)$. Given a profile $R^N \in W^N$ and any $x, y \in X$, $x D(R^N) y$ if there is a coalition $K \in V$ such that x is strictly preferred to y by all members of K. $C(G, X, R^N)$ consists of alternatives not dominated by any other alternatives.

We have already associated a matrix-form game to an SDF. We shall now associate a simple game to an SDF (Peleg 1978). A coalition K is winning with respect to SDF f if for all $R^N \in W^N$ and for all $x \in X$ the fact that x is strictly preferred to y by all members of K implies that $f(R^N) = x$. Since the set of all winning coalitions is denoted by V we can now associate to any SCF f a game $G_f = (N, V)$.

A simple game G is proper if $K \in V$ implies $N \setminus K \notin V$. G is strong if $K \notin V$ and $K \subset N$ imply that $N \setminus K \in V$. Finally, G is weak if the intersection of all coalitions in V is non-empty. In other words, G is weak when there are individuals − called vetoers − that have to be present in all winning coalitions.

6.2.2   Some results

Much of the literature on voting games deals with equilibria and other solution concepts. One of the most important solution concepts is that of a core, which was defined above. From the social choice theory perspective a core is a generalization of a Condorcet winner in majority voting games. As was pointed out above, a core is the set of non-dominated alternatives. It is of considerable interest to know when a non-empty core exists. Fortunately, a complete answer to this problem has been given by Nakamura (1979) and Peleg (1984). This answer has direct bearing upon the best known result in the voting game genre, namely the theorem of Gibbard (1973) and Satterthwaite (1975).

As was discussed above, the theorem states that all universal and non-trivial SCFs defined for alternative sets with at least three elements are either manipulable or dictatorial. Universality of an SCF f means that f be defined for all $R^N \in W^N$, while non-triviality requires that for any $x \in X$ there is at least one preference profile L^N such that $f(L^N) = x$. An SCF is dictatorial if there exists an individual d such that for any $x \in X$ there is an R_d so that $f(R_d, R^{N-\{i\}}) = x$ for all $R^{N-\{i\}} \in W^{N-\{i\}}$.

The Gibbard−Satterthwaite theorem turns out to be a corollary of a more general theorem of Ishikawa and Nakamura (1979). Let us first define the Nakamura number of a simple game $G = (N, V)$ as

$N(G) = \min M$, where

$$M = \{card(K) \mid K \subset V, \cap S \mid S \in K\} = \emptyset.$$

Here $card(K)$ denotes the number of elements in set K.

The Nakamura number, thus, is the minimum number of winning coalitions with an empty intersection. In simple majority voting games the Nakamura number is 3. In fact, Ishikawa and Nakamura (1979) have shown that for practically all decisive and intuitively democratic − namely proper, strong and not weak − simple games the Nakamura number equals 3.

The following theorem is proven by Nakamura (1979). Let G and X be defined as above. If $C(G, X, R^N)$ is non-empty, then either there exists a set of veto players or the Nakamura number of G is strictly larger than $card(X)$. The result from which the Gibbard−Satterthwaite theorem follows as a corollary can now be stated. Let $card(X)$ be at least 3 and let f be an SCF. Moreover, let G_f be an associated simple game. If now f is coalitionally strategy-proof then either one of the two following conditions, but not both, holds:

- G_f is proper, strong and $N(G_f) > card(X)$.
- f is dictatorial.

In view of Nakamura's result just cited, the first condition cannot hold if $card(X)$ is larger than 2. Thus the second must hold. In other words, f is either coalitionally and thus, by Ishikawa and Nakamura's (1979) lemma, individually manipulable or dictatorial.

Another result of considerable importance for voting games is related to the existence of a non-empty core. As was pointed out above, this result has also been achieved by Nakamura (1979). Stated in somewhat intuitive fashion, the theorem says that when G is not weak, the core is always non-empty if and only if the number of alternatives is strictly less than the Nakamura number of the associated simple game. In essence this result thus does away with the possibility of the existence of games with no vetoers that would always have non-empty cores.

One feature in the social choice literature that has been a source of some confusion is the tendency of different authors to resort to different basic assumptions. Thus, some authors start from assuming that the voters have complete and transitive binary preference relations. However, sometimes a more specific assumption, namely that of a complete, transitive and antisymmetric relation is made even by the same authors (see, for example, Peleg 1984 and Moulin 1983; also Blair 1981). Also different rules are being studied. As we observed above, in his classic

Group I	Group II	Group III
A	B	C
B	A	B
C	C	A

Table 6.3: Zermelo's Algorithm in Group Choice

treatise Arrow (1963) was focusing on social welfare functions, while Gibbard (1973) and Satterthwaite (1975) obtained their result concerning SDFs. The most recent literature typically studies SCCs.

In parallel with the studies on strategy-proofness of SCFs another stream of research has focused on specific voting institutions. This tradition is motivated by Farquharson's work (Farquharson 1969). In the following section we shall discuss the main results achieved in this research.

6.3 SOPHISTICATED VOTING AND AGENDA INSTITUTIONS

Farquharson's example from the Roman Senate discussed in Section 6.1 is based on an agenda procedure. To predict what will be the outcome of the procedure under a given preference profile one needs to know both the agenda and the voting strategies. The latter are, of course, partly determined by the preferences, but also by the type of information available to the voters. As was pointed out above, two main assumptions about voting strategies are sincere and strategic (or sophisticated). To find out the outcomes ensuing from strategic voting under complete information has been a major preoccupation of many theorists since the publication of Farquharson's work. As was pointed out above, Farquharson's method proceeds through elimination of dominated voting strategies. This method sometimes leads to unique predictions. However, when the set of alternatives is even moderately large, the method is extremely cumbersome. An easier method, suggested by McKelvey and Niemi (1978), is based on Zermelo's algorithm. It was discussed in the preceding, but we shall now demonstrate its application in the voting context. The preference profile is given in Table 6.3. Suppose that the amendment procedure is used with the following agenda:

1. A versus B,
2. the winner of the previous stage versus C.

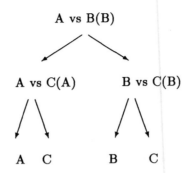

Figure 6.1: Game Tree of the Group Choice Example

The procedure can be represented as in Figure 6.1. For each node of the tree a strategic equivalent is determined by proceeding from the final outcomes upwards towards the beginning of the tree. The strategic equivalent of the node A vs C is the outcome which would win this comparison by a simple majority if voting were sincere. Obviously the winner would be A. But why assume that the voting is sincere? The reason is simple: if this node is ever reached, it is rational for the players to vote according to their sincere preferences since to vote otherwise could only harm them if it changed anything. Thus, in the nodes preceding the final outcomes, we can rest assured that the voters will vote according to their preferences. In fact, we can replace this node with outcome A since there is no reason to expect any other outcome to ensue from this node. By a similar reasoning we can replace the node B vs C with B. The strategic equivalents of each node are written in parentheses in Figure 6.1.

It turns out that the strategic equivalent of the root of the tree, that is, the first pairwise comparison, is B, the Condorcet winner. This is no accident: the Condorcet winner always coincides with the outcome of sophisticated voting, that is, the strategic equivalent of the voting game tree.

A general algorithm for finding the set of sophisticated voting outcomes has been devised by Shepsle and Weingast (1984). It is defined for any preference profile and agenda. Given an agenda, it constructs its strategic equivalent agenda. The example in Table 6.4 demonstrates this. The agenda is:

1. A versus B
2. the winner of the previous stage versus C

persons 1,2	persons 3−5	persons 6,7
A	B	C
C	D	D
B	A	A
D	C	B

Table 6.4: A Demonstration of the Shepsle−Weingast Algorithm

3. the winner of the previous stage versus D.

The strategic equivalent agenda is formed as follows:

- the last element of the constructed agenda is the same as in the original agenda. In our example the last element is D.
- the ith element is identical with the ith element of the original agenda if and only if it defeats the later elements of the agenda with a majority of votes according to the preference profile. Otherwise the ith element is identical with the $i + 1$th element of the original agenda.

In the example C defeats D. Hence the penultimate element of the strategic equivalent agenda is C. The second element is also C since neither A nor B defeats both C and D. Thus, the first element is also C. The entire strategic equivalent agenda is thus: C,C,C,D.

Shepsle and Weingast show that the first element of the strategic equivalent agenda coincides with the outcome of sophisticated voting under the given agenda. Thus, their algorithm enables us to find out the sophisticated voting outcomes without resorting to the voting game tree and backwards induction.

6.3.1 Solution concepts

The Pareto set (PS, for short) consists of those alternatives that are unanimously preferred to any alternative outside the set. It is typically a very large set. In the four preceding examples the Pareto set contains all alternatives. The criterion of unanimity is the reason for its indecisiveness. The other solution concepts are based on pairwise comparison and typically resort to the majority winning criterion.

The top cycle set (TC) is based on a generalization of the Condorcet winner concept. TC consists of the smallest set of alternatives such that all members of the set defeat each alternative outside the set. Obviously, if a Condorcet winner exists, it is the only alternative in TC.

The concept of covering appears frequently in the graph-theoretical literature. It was introduced to the theory of voting games by Miller

(1977; 1980; 1995). The binary relation C (covering) over the set of alternatives X is defined as follows: xCy if and only if x defeats y and everything that y defeats. In voting games the criterion of defeating is typically that of a simple majority in pairwise comparison. The uncovered set (UC, for brevity) consists of alternatives that are covered by no other alternative.

Obviously, when a Condorcet winner exists, it is the sole element of UC. On a closer inspection it turns out that all sophisticated voting outcomes are located in UC, that is, no matter which particular agenda is being used, the outcomes of sophisticated voting are in UC. The converse, however, is not necessarily true. In other words, there may be alternatives in UC that are not outcomes of sophisticated voting under any agenda. The set UC is thus somewhat too large if one wants to characterize exactly the outcomes ensuing from sophisticated voting. The question hence arises as to whether it is possible to find an algorithm that would generate all sophisticated voting outcomes and only them.

The question has been answered affirmatively by Banks (1985; see also Miller et al. 1990a; 1990b). Given an alternative set X and an alternative $x \in X$, define a Banks chain starting from x so that x is the end point of the chain if there is no $y \in X$ that defeats x. If such a y exists, then y is the end point of the chain starting from x, unless there exists an alternative, say z, that defeats all the preceding elements, that is, x and y in the chain. Eventually an end point will be reached for the Banks chain starting from any alternative. The Banks set (BS) consists of the end points of all Banks chains. The set coincides exactly with the set of sophisticated voting outcomes. We thus have an algorithm for determining the set of sophisticated voting outcomes for amendment-type voting games.

6.3.2   Relationships between Solutions

Since the Banks set contains precisely all sophisticated voting outcomes, it is of some interest to find out how this set is located *vis-à-vis* the other solutions defined above. First of all, when a Condorcet winner exists, BS, TC and UC all coincide with it and are contained in PS. PS, on the other hand, typically contains other alternatives in addition to the Condorcet winner.

When no Condorcet winner exists, the situation becomes more complicated. BS is still within PS and within TC. There is no necessary set inclusion relation between the latter two sets, that is, TC may contain elements not in PS and *vice versa*. However, BS and UC are both within

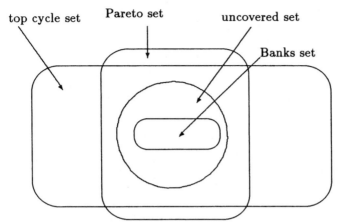

Figure 6.2: Relationships between Solutions

the intersection of PS and TC (Nurmi 1995). In the absence of a Condorcet winner, UC and BS are identical for small alternative sets. Once the number of alternatives gets larger than six, BS is always a subset of UC (not necessarily proper, though)(Moulin 1986). Figure 6.2 depicts the relationships between these solutions.

Further reading: Miller (1995); Moulin (1986); Nurmi (1995).

6.4 IMPLEMENTATION

The set A of outcomes is implementable if there exists a game form such that the outcomes in A are identical with the equilibria of the game form. A game form is simply the outcome function f, that is, a mapping from strategy n-tuples to outcomes. In the literature the outcome set A is typically defined by means of SCC. Thus, the SCC F is said to be implemented by g if for any preference profile $R^N \in W^N$: $F(R^N) = eq\ g(f, R^N)$, where eq denotes equilibria of a game. Note that each preference profile in g is associated with a set of equilibria that result from the players' choice of equilibrium strategies. These may or may not involve reporting their true preferences.

In the implementation literature several types of equilibria have been discussed, most notably Nash and dominant strategy equilibria. Usually the above requirement that the same game form gives the outcomes of SCC as equilibria is somewhat relaxed so that the requirement is that the outcomes of the game form be a subset of the values of the SCC.

Sometimes the requirement that only one game form be used is also dispensed with (Maskin 1985). Theoretically the most important types of implementability are (i) those in which the equilibria have to form a subset of the SCC values, and (ii) those in which the set of equilibria is required to coincide with the set of SCC values. Case (i) is that of weak implementation, while case (ii) defines ordinary implementation. Repeating the definitions once more we say that g weakly implements F if for all $R^N \in W^N$, eq $g(f, R^N) \subset F(R^N)$. On the other hand, g implements F if for all $R^N \in W^N$, eq $g(f, R^N) = F(R^N)$.

Since SCCs can be associated with voting procedures, it is obvious that the results on implementation can be directly applied to voting. Game forms are precisely the angle from which one looks at voting when it is viewed as a game of strategy.

6.4.1   Implementation in Nash equilibrium

The fundamental result on implementation in Nash equilibrium is due to Maskin (1985). It gives two sufficient conditions for Nash implementation. The first is called weak no-veto power. An SCC F satisfies weak no-veto power if for all $R^N \in W^N$ and $x \in X$: $x \in F(R^N)$ whenever there exists an individual i such that for all other individuals j and alternatives $y \in X$: xR_jy. Thus, any alternative is chosen by F if it is ranked first by all individuals save possibly one.

The second condition is called Maskin monotonicity in contradistinction from 'ordinary' monotonicity. An SCC F is Maskin monotonic if for all $R^N, S^N \in W^N$ and for all $x \in X$: $x \in F(S^N)$ whenever (1) $x \in F(R^N)$ and (2) for all $y \in X$ and all individuals i: xR_iy implies xS_iy. This condition thus requires that if an alternative x is chosen under profile R and profile S is formed from R so that x is ranked at least as high as y in S than in R for all y, then x should be in the choice set in S as well. This notion should be kept distinct from that of monotonicity discussed above. As will be recalled, the latter requires that S be formed from R so that x's position is no worse in S than in R, *ceteris paribus*, that is, keeping the other preferences fixed.

An example of an SCC that is monotonic but not Maskin-monotonic is the Borda count. In the following example (Table 6.5) the Borda choice set is $\{A, B, C\}$. Suppose now that all five voters in the middle group change their preference between A and B (that is, from CBA to CAB) and that one voter of the right-most group changes her preference between A and C (that is, from BAC to BCA). Clearly these changes result in a new profile where C's rank with respect to that of any other alternative has not deteriorated. In fact one voter has changed her mind

5 voters	5 voters	5 voters
A	C	B
C	B	A
B	A	C

Table 6.5: Borda Count and Maskin Monotonicity

strictly in favour of *C* *vis-à-vis* *A*. Yet in the new profile *A* is the sole Borda winner, showing that the Borda count is not Maskin-monotonic. Yet it is obviously monotonic.

Maskin's result is that Maskin monotonicity and weak veto power are sufficient for implementation in Nash equilibrium. The former condition is also necessary for games with at least three players. An analogous condition for Nash implementability, called essential monotonicity, has been discussed by Danilov (1992). Very few systems satisfy Maskin monotonicity or essential monotonicity. This means, then, that very few systems are implementable in Nash equilibria. On the other hand, we have seen that all Nash equilibria are not always plausible or there are simply too many of them. Hence it may be downright unreasonable to insist that all values of an SCC be Nash equilibria. It could well be that some proper subset of Nash equilibria gives a more plausible goal to strive for. Therefore, it is worthwhile to ask if systems exist that are implementable in other types of equilibria.

6.4.2 Subgame-perfect implementation

If the Nash equilibrium is discarded as too liberal, then the most obvious candidate for the equilibrium notion is that of subgame-perfect equilibrium. Results on implementability in subgame-perfect equilibria have been achieved by Moore and Repullo (1988) as well as by Abreu and Sen (1990). These authors investigate the necessary and sufficient conditions for implementability in subgame-perfect equilibria. Since the subgame-perfect equilibrium concept is defined with respect to extensive-form games, we shall start with the definition.

Definition 6.1 *An extensive-form game g is an array $g = (T, >, D, \delta, w)$, where T is the set of nodes of the game tree, $>$ is a partial ordering of precedence, D is the set of decisions, δ indicates for each non-initial node the decision that led to it from its predecessor node and w is a function that indicates for each terminal node the outcome ensuing from it.*

The strategy set of each player i is denoted by S^i. A strategy n-tuple $(s^1, \ldots, s^n)$ is a subgame-perfect equilibrium if for each $i \in N$, for any

$s^i \in S^i$ and for each non-terminal node t:

$$w((s^i, s^{-i}), t) R_i w((s^{i'}, s^{-i}), t)$$

where $s^i \in S^i$ and s^{-i} denotes the strategy $n - 1$-tuple of all the other players except i. In other words, a strategy n-tuple is a subgame-perfect equilibrium if every player is choosing the best reply to the other players' choices in every node preceding the terminal one. The definition of Nash equilibria of an extensive-form game is obtained by replacing t with t_0, the initial node, that is, the n-tuple of strategies is a Nash equilibrium if every player is choosing the best response to the others' choices at the beginning node of the game. As was pointed out above, the subgame-perfect equilibria are a subset (refinement) of Nash equilibria.

The necessary condition − called C − for subgame-perfect implementation as defined by Moore and Repullo is somewhat complicated, although as a requirement it is very mild. C plays a much more modest role in subgame perfection than Maskin monotonicity plays in Nash implementation. The content of C bears some family resemblance to Maskin monotonicity, though. Consider an SCC denoted by F and two profiles, P and Q. Assume that an alternative $x \in F(P) \backslash F(Q)$. In other words, x belongs to the choice set of F when the profile is P but not when the profile is Q. Condition C is the requirement concerning the existence of a subset B of alternatives with the following property: under these circumstances there exists a sequence of alternatives $\{a_0 = x, \ldots, a_m\}$ all in B such that for each $k = 0, \ldots, m - 1$ there is a player i_k who in profile P regards the $k - 1$th alternative in the sequence at least as good as the kth alternative and a player j whose preference between a_m and a_{m+1} is $a_m R_j a_{m+1}$ in profile P and $a_{m+1} R_j a_m$ in Q. So, B consists of alternatives such that in P (where x belong to the winners) they can be ordered with x at the top so that for each pair of alternatives there is a voter whose preference with regard to this pair coincides with the order in the sequence. Moreover, in Q there has to be a player who switches her opinion between the last two alternatives in B.

Intuitively C amounts to requiring that for x to vanish from the choice set when P is replaced with Q it is necessary that some individuals change their minds. However, in contradistinction to Maskin monotonicity, which also requires preference changes involving x, this change is not necessarily related to x. While Maskin monotonicity requires that changes which do not affect x's ordinal position *vis-à-vis* other alternatives do not exclude x from the choice set (if it belongs to it in the first place), condition C requires that an alternative be excluded from a choice set just in case some people change their minds about the ranking of some alternatives. As Abreu and Sen point out, condition

voter 1	voter 2	voter 3
A	C	B
B	A	C
C	B	A

Table 6.6: Abreu and Sen's Example

C is very mild indeed. In fact, it is satisfied by all SCCs satisfying the Pareto criterion provided that the preferences are all strict.

Although necessary for subgame-perfect implementation, condition C is not sufficient. Abreu and Sen show that C can be made almost sufficient by adding the requirement that the preference changes be connected to x in a specific way. C augmented with this additional requirement is called condition α. No-veto power and α together imply subgame-perfect implementability (Abreu and Sen 1990).

To get an idea of the nature of condition α let us borrow an example from Abreu and Sen (1990). (See Table 6.6).

Let us call this profile P and let us call Q the profile obtained from P by changing voter 3's preference between B and C so that her order is CBA. The SCC called majority voting chooses the Condorcet winner whenever one exists. Otherwise it chooses the entire alternative set. Clearly, then, $F(P) = \{A, B, C\}$ and $F(Q) = \{C\}$. Since the preference change from P to Q does not involve deterioration of A's rank *vis-à-vis* any other alternative, Maskin monotonicity would require that $A \in F(Q)$. Since this is not the case, the majority voting is not Maskin monotonic. Consequently, it is not Nash-implementable. It is, however, implementable in subgame-perfect equilibria. The reason for this conclusion is that it satisfies α and the no-veto power condition. The former condition amounts to finding a sequence of alternatives satisfying condition C and, moreover, having the property that a_k is not ranked first in Q by i_k for any $k = 0, \ldots, m$. Additionally the sequence has to be such that if a_{m+1} is ranked first in Q by all players except i_m, then either $m = 0$ or $i_{m-1} \neq i_m$. In the above example $A \in F(P) \setminus F(Q)$. The sequences of alternatives and voters required by α are: $a_0 = A, a_1 = B, a_2 = C$ and $i_0 = 2, i_1 = 3$. Also $B \in F(P) \setminus F(Q)$. In B's case the corresponding sequences are: $a_0 = B, a_1 = C$ and $i_0 = 3$. Since majority voting obviously satisfies no-veto power condition, it is implementable in subgame-perfect equilibria.

In a recent paper Sertel and Yilmaz (1997) show that the SCC called majoritarian compromise is implementable in subgame-perfect equilibria. The majoritarian compromise procedure is defined for any preference profile P and $F(P)$ is computed as follows. First one determines

voter 1	voter 2	voter 3
A	C	B
C	A	C
B	B	A

Table 6.7: Modified Abreu and Sen's Example

voter 1	voter 2	voter 3
A	C	B
B	A	C
C	B	A

Table 6.8: Majoritarian Compromise and Maskin Monotonicity

the number of first ranks for any alternative. If this number exceeds 50 per cent of the voters for some alternative, then this alternative is the winner. Otherwise, one computes for each alternative the number of first and second ranks. If this number exceeds the above threshold, then the corresponding alternative is declared the winner. Otherwise, one continues computing the third ranks and so on. Eventually some alternative is found to exceed the 50 per cent limit. The first such alternative is the winner. Suppose now that counting ranks $1, \ldots, k$ gives no winner, while counting ranks $1, \ldots, k+1$ gives several alternatives that exceed the threshold. According to the majoritarian compromise principle the winner is the alternative with most ranks $1, \ldots, k+1$. Thus the ties are broken in a 'natural' way. For illustration, consider a slightly modified version of Abreu and Sen's example (Table 6.7). Counting first ranks yields no winner, while counting the first and second ranks indicates that A and C have both exceeded the 50 per cent threshold. However, C is ranked first or second by all voters, while A is ranked first or second by only two of them. Hence, C is the winner according to the majoritarian compromise.

The majoritarian compromise fails on Maskin monotonicity. Table 6.8 is a slight modification of Sertel and Yilmaz's example.

Denoting this profile P, we get the majoritarian compromise choice set $F(P) = \{A, B, C\}$. Suppose now that voter 3 changes her mind about the preference between B and C so that in the new profile Q her preference order is CBA. No other changes are made. Clearly voter 1's change of mind has not deteriorated A's rank *vis-à-vis* any other alternative. Yet $F(Q) = \{C\}$. Hence majoritarian compromise does not satisfy Maskin monotonicity. Consequently, it is not implementable in

Nash equilibria.

It is, however, implementable in subgame-perfect equilibria (Sertel and Yilmaz 1997). Other similar systems are Copeland's SCC, as well as the majority SCC as was pointed out above (Abreu and Sen 1990).

Although social choice and mechanism design theories are seemingly dealing with closely interrelated issues, the specialization has led to a situation in which the conceptual apparatus and results of one field are not easily translatable into the language of the other. When one looks at voting as a game strategy, the results of both theories are apparently relevant. It takes some effort, however, to find out what these results imply in terms of voting procedures or other practical ways of aggregating opinions or performance criteria.

The most important consideration in interpreting the results discussed above calls for a scrutiny of the information that the voters are assumed to possess. The safest equilibrium concept is that resulting from dominant strategies. If each voter always has a unique dominant voting strategy and this always coincides with the values of the SCC under investigation, then the implementation problem is solved. Farquharson's work is an attempt to generalize this type of reasoning by successively eliminating dominated voting strategies. This work led to Moulin's important results concerning SCCs which are implementable in sophisticated voting equilibria (Moulin 1979). The assumptions concerning the distribution of information among voters are relatively mild: to eliminate dominated voting strategies one does not have to know the strategy choices actually made by the others.

The literature on Nash implementation, on the other hand, rests on the assumption that unilateral deviation from a strategy n-tuple is not beneficial for the deviator. It is well known that the class of Nash equilibria is often so large that it includes downright implausible strategy combinations. As was discussed above, the necessary condition for Nash implementability is Maskin monotonicity. This condition is very rare among voting systems. Thus the conclusion of Gibbard and Satterthwaite about the manipulability of practically all non-dictatorial SCFs is generalizable to SCCs. At the same time, the intuitive weaknesses of the Nash equilibrium concept suggest that one should look for a more plausible equilibrium concept. Subgame-perfect equilibrium is one possible candidate. The results on implementability in subgame-perfect equilibria suggest that several procedures satisfy the sufficient conditions for this type of implementability while failing on Maskin monotonicity.

The work on more general agenda structures has made considerable progress but the most complete results discussed above pertain to amendment-type agendas (for results on other types, see Banks 1989;

Srivastava and Trick 1996). For our purposes it is sufficient to observe that the results seem to suggest that non-myopic behaviour effectively counteracts the chaotic tendencies of the simple majority rule in delineating typically a fairly small subset of alternatives within which the outcomes will be found under sophisticated voting. Thus, even in the absence of the core, the majority voting is less arbitrary under sophisticated voting than under myopic voting.

Further reading: Maskin (1985); Moore and Repullo (1988); Abreu and Sen (1990).

6.5   BIBLIOGRAPHICAL REMARKS

The first systematic work on strategic voting was the article published in 1961 by Michael Dummett and Robin Farquharson (Dummett and Farquharson 1961). It was followed by the latter's seminal monograph some years later (Farquharson 1969). After that the growth in the literature on this subject has been rapid. Particularly important contributions are Moulin's, Ishikawa and Nakamura's as well as Peleg's works (Moulin 1983; Ishikawa and Nakamura 1979; Peleg 1984). The article by McKelvey and Niemi (1978) is very important for establishing the link between games and social choice. It is a particularly useful step towards characterizing sophisticated voting outcomes. A superset of those outcomes was first outlined by Miller and exact boundaries were finally drawn by Banks (Miller 1980; Banks 1985; Miller et al. 1990a; 1990b). The relationships between various tournament solutions concepts (such as the set of Copeland winners and Banks set) are analysed by Moulin, Miller, Grofman and Feld (Moulin 1986; Miller et al. 1989). Miller's book is a comprehensive exposition of agenda institutions (Miller 1995).

A relatively non-techical introduction to implementability and mechanism design is Feldman's textbook (Feldman 1980). The great names of this branch of research are Theodore Groves, Leonid Hurwicz, John Ledyard, Jacob Marschak and Roy Radner (Groves and Ledyard 1977; Hurwicz 1972; 1979; Ledyard 1995; Marschak and Radner 1972). The basic results on Nash implementation are reported in Eric Maskin's article (Maskin 1985). Works on subgame-perfect implementability are more recent.

7 Institutional Design and the European Union

Thus far we have not touched upon the cooperative-game applications to institutional design. The recent developments in the European Union offer a suitable field for applying to institutional design (or reform) the theoretical insights stemming from these games. Also other transnational organizations, for example the United Nations, are under considerable pressure towards institutional reform. Similar considerations would undoubtedly be applicable in those cases as well. In this chapter we shall, however, focus on the European Union.[1]

A considerable amount of work has been put into evaluating the likely consequences of various proposals for institutional reform. What the players of the EU game are particularly interested in is their share of influence after various reforms. Therefore, the issue of how to measure this influence is of paramount importance.

7.1 VOTING POWER INDICES

Proportional election systems are designed to achieve a reasonably good match between the power relationships of various political groups in the electorate and the power relationships of the parties represented in parliament. Typically this is achieved by distributing the parliamentary seats to parties in proportion to their support in the electorate: with a

---

[1] In this chapter the abbreviation EU stands for the European Union rather than expected utility as was the case earlier in the book.

support of x per cent of the electorate, a party is entitled to roughly x per cent of the seats. This approach completely overlooks the fact that in parliaments the legislative decisions are made using various decision rules, such as simple majority, 3/4 majority and so on. In a two-party system a party having 40 per cent of the seats in parliament can hardly expect to have 40 per cent of the power over legislative outcomes if the other party has 60 per cent of the seats.

7.1.1   Definitions and an example

Intuitively speaking, various decision rules distribute power differently in voting bodies even though the number of votes of individuals or groups does not vary. Consider a three-party parliament consisting of 100 seats. Party A_1 has 40 seats, party A_2 has 35 seats and party A_3 has 25 seats. Let us now ask what is the distribution of influence over legislative outcomes, that is the distribution of voting power, among the parties?

Over the past decades several different ways of answering this question have been proposed. They are all based on different ways of measuring voting power. The oldest one is the Shapley–Shubik index (Shapley and Shubik 1954).

Definition 7.1 *Shapley–Shubik index: power equals the relative number of pivotal positions assuming that all player permutations are equally likely.*

A permutation is an ordered sequence of objects. Two permutations can be formed of two objects, six of three objects, and so on. In general, the number of permutations of n objects is $n! = n \cdot (n-1) \cdot (n-2) \cdot \ldots \cdot 2 \cdot 1$. A player is pivotal in a permutation if he together with all the players on his left side form a group (coalition) that has at least as many votes as the decision rule, while the group that consists of the voters on his left side without him has strictly fewer votes than the decision rule.

Another power index is based on coalitions rather than permutations (Banzhaf 1965). Coalitions are simply subsets of players.

Definition 7.2 *Banzhaf index:*[2] *power equals relative number of critical presences ('swings').*

A player is critically present in a coalition if the members of this coalition have altogether at least as many votes as the decision rule, but fewer than the decision rule if the player leaves the coalition.

---

[2]It would perhaps be more appropriate to call this index the Penrose index since it was proposed by L.S. Penrose about twenty years before Banzhaf — undoubtedly independently — published his article (Penrose 1946; Felsenthal and Machover 1997; Nurmi 1997c). I shall, however, stick to the prevailing terminology here.

Since the power indices are solution concepts of n-person cooperative game theory, it is useful to outline briefly the basic concepts of the theory. The n-person game is typically defined by means of a characteristic function which is a mapping from all subsets of players into some subset of real numbers, that is

$$v : 2^n \Rightarrow A \subseteq \mathbf{R}$$

where 2^n denotes all coalitions (subsets) that can be formed of n players and $\mathbf{R}$ the set of real numbers. The set of all players is denoted by N and the decision rule by k. For a subset S of N, $v(S)$ indicates the value of S if it is formed. This function is one of the 'givens' of the theory, that is, its properties are typically assumed rather than derived.

The power indices are defined for so-called simple games. These are games in which any coalition is either winning or non-winning. Accordingly, the characteristic function has only two values, 0 and 1. The former value characterizes non-winning coalitions and the latter winning ones. A simple n-person game can be defined as a pair (N, W) where W is the set of all winning coalitions. We denote by $w(S)$ the number of votes of the members of coalition S. The set W is thus determined by the following:

$$S \in W \text{ iff } w(S) \geq k.$$

In other words, a coalition is winning iff it has at least as many votes as the decision rule.

A pair $(S, S \setminus i)$ denotes a critical presence or swing of player i iff S is winning and $S \setminus i$ (that is, S with i excluded) is non-winning. Both the Shapley–Shubik and Banzhaf indices utilize the notion of swing in defining the *a priori* voting power of players. We use the previous example of a 100-member parliament to illustrate these concepts.

Instead of individual MPs consider parties as the players of the game. Each row of Table 7.1 indicates one permutation of the three players. The second column gives the pivotal party in the corresponding permutation assuming that the decision rule is 51. The third column indicates the pivotal party for the decision rule 67.

The Shapley–Shubik index value of a player is the number of times she is pivotal in all permutations divided by the number of all permutations. Since in each permutation one and only one player is pivotal, the sum of Shapley–Shubik index values of all players is necessarily 1. Denoting the Shapley–Shubik index values of A_1, A_2 and A_3 by ϕ_1, ϕ_2 and ϕ_3, we get

$$\phi_1 = \phi_2 = \phi_3 = 2/6 = 1/3,$$

permutations	pivotal party for $k = 51$	pivotal party for $k = 67$
$A_1 A_2 A_3$	A_2	A_2
$A_1 A_3 A_2$	A_3	A_2
$A_2 A_1 A_3$	A_1	A_1
$A_2 A_3 A_1$	A_3	A_1
$A_3 A_1 A_2$	A_1	A_2
$A_3 A_2 A_1$	A_2	A_1

Table 7.1: Player Permutations in a Fictitious Parliament

for $k = 51$, and

$$\phi_1 = \phi_2 = 3/6 = 1/2, \phi_3 = 0,$$

for $k = 67$.

For computing the Banzhaf index values we first list the swings of each player when $k = 51$.

Swings of A_1: $\{A_1 \cup A_2, A_2\}, \{A_1 \cup A_3, A_3\},$

Swings of A_2: $\{A_1 \cup A_2, A_1\}, \{A_2 \cup A_3, A_3\},$

Swings of A_3: $\{A_1 \cup A_3, A_1\}, \{A_2 \cup A_3, A_2\}$

The (standardized) Banzhaf index value of player $A_i, (i = 1, 2, 3)$, denoted β_i, is calculated by dividing the number of his swings by the sum of swings of all players. Thus, for $k = 51$,

$$\beta_1 = \beta_2 = \beta_3 = 2/6 = 1/3.$$

When $k = 67$, there are no swings for A_3. There are only two coalitions that are winning, the 'grand' coalition consisting of all players and the coalition formed by A_1 and A_2. Both A_1 and A_2 have swings in these coalitions. Consequently, for $K = 67$

$$\beta_1 = \beta_2 = 2/4 = 1/2, \beta_3 = 0.$$

Thus both power indices give the same distribution of *a priori* voting power for both decision rules. This is, however, not always the case, as we shall see later on. The power indices are called *a priori* ones to emphasize their theoretical nature: they make simplifying assumptions about the coalition formation process and utilize no other information than the vote distribution and the decision rule in computing the values. Therefore they cannot be expected to reflect such considerations as coalition preferences of players, ideological proximity and so on.

Having seen how the indices are computed in a simple example, we can now state the formal definitions of the power indices. It is perhaps helpful to notice that both the two Banzhaf indices and the Shapley–Shubik one add up the contributions a player makes to all possible coalitions, the difference being that in the latter index these contributions are weighted and in the former unweighted. In the following definitions s denotes the number of players in coalition S and n the number of players in N, that is, the entire player set.

The Shapley–Shubik index value of i:

$$\phi_i = \sum_{S^{\bullet} \subseteq N} \underbrace{\frac{(s-1)!(n-s)!}{n!}}_{\text{the weight}} \underbrace{[v(S^*) - v(S^* - \{i\})]}_{i\text{'s contribution}}$$

The (standardized) Banzhaf index value of i:

$$\beta_i = \frac{\sum_{S^{\bullet} \subseteq N}[v(S^*) - v(S^* - \{i\})]}{\sum_i \sum_{S^{\bullet} \subseteq N}[v(S^*) - v(S^* - \{i\})]}$$

In addition to the standardized index there is also another version of the Banzhaf index, namely the absolute one:

$$\beta_i' = \frac{\sum_{S^{\bullet} \subseteq N}[v(S^*) - v(S^* - \{i\})]}{2^{n-1}}$$

The absolute Banzhaf index differs from the standardized one in the denominator; in place of the total number of swings the absolute index has 2^{n-1}. This means that the absolute Banzhaf index values do not in general add to unity.

Although the above three indices are by far the most widely applied ones in institutional design, for example in discussions concerning the effects of various modifications of decision-making rules to the power distribution, they are not the only ones. Two other indices are defined in the following. These are based on minimal winning coalitions which are coalitions that consist of essential members only. In other words, the removal of *any* player from a minimal winning coalition results in a losing coalition. This is be distinguished from winning coalitions. In the example of Table 7.1 above, when $k = 67$, $A_1 A_2 A_3$ is a winning coalition, but not a minimal winning one since A_3 can be removed from it without the coalition becoming non-winning. In other words, in minimal winning coalitions all players have a swing. This is not necessarily the case in all winning coalitions. The two indices below take account of the minimal winning coalitions only.

Denote by $\mathcal{M}$ the set of all minimal winning coalitions. The Holler index value of player i, denoted H_i, is calculated as follows:

$$H_i = \frac{\Sigma_{S* \subseteq \mathcal{M}}[v(S*) - v(S * \setminus \{i\})]}{\Sigma_{j \in N} \Sigma_{S* \subseteq \mathcal{M}}[v(S*) - v(S * \setminus \{j\})]}.$$

The Deegan-Packel (DP) index value of actor i, denoted J_i, is computed as follows:

$$DP_i = \frac{\Sigma_{S* \subseteq \mathcal{M}} 1/s[v(S*) - v(S * \setminus \{i\})]}{\Sigma_{j \in N} \Sigma_{S* \subseteq \mathcal{M}} 1/s[v(S*) - v(S * \setminus \{j\})]}.$$

Here s denotes the number of members in the minimal winning coalition $S*$.

The Holler index value of a player equals the number of his swings in minimal winning coalitions divided by the swings of all players in minimal winning coalitions. Thus it is essentially an application of the intuition underlying the Banzhaf index to minimal winning coalitions. The Deegan-Packel index also counts swings in minimal winning coalitions, but assigns each one of them a weight which equals the inverse of the number of players in the corresponding coalition. These weighted swings are then divided by the sum of all weighted swings so as to end up with a standardized measure.

Further reading: Banzhaf (1965); Brams (1975); Deegan and Packel (1982); Holler (1982a); Holler and Packel (1983); Johnston (1995); Shapley and Shubik (1954).

7.1.2   Which index is right?

In view of the multiplicity of the indices one is led to ask which one, if any, of them is most appropriate for measuring the voting power of players. Before attempting to answer this, let us be more specific about what kind of power we are interested in. There are many kinds of power, some highly contextual, issue-dependent and difficult to measure. For example, a popular television series may affect the way millions of persons arrange their daily routines. The news media influence the way people orient themselves in political issues. These influences are difficult to measure although few would deny that they exist. Power indices deal with influences that the actors exert upon decision outcomes, that is, those decisions that are made by the collective body of which they are part.

On the basis of this preliminary characterization of what power indices measure, we are in a better position to answer the question of the most appropriate index. The most appropriate index is one which is based on correct assumptions with regard to the way outcomes are generated. Underlying each outcome is a set of assumptions about how the outcomes are reached. If one knows the decision-making body in which the index is to be applied, then one should compare those assumptions with the principles according to which outcomes are reached in that body. For example, if only minimal winning coalitions are expected to form in the collective body, then the appropriate indices are of either the Holler or Deegan-Packel variety. If, on the other hand, any winning coalition is about equally likely to dictate the outcomes, then the Banzhaf indices should be applied. Or, if any attitudinal dimension − that is, an ordering of voters from the strongest supporter to the fiercest opponent − is assumed equally likely, then the Shapley−Shubik index would do the trick.

Since the indices are typically used in contexts where the issues to be decided are not known, it is often impossible to say what kind of coalitional mechanism will eventually be at work in the body. This is precisely why the indices are called *a priori* measures of power. They should not be confused with 'real' power, that is, the actual influence on the outcomes which can only be assessed *ex post*. One should interpret power index values as telling what influence one could expect an actor with given resources to have on the outcomes if:

- the decision rule is given,
- the distribution of resources over the other players is given, and
- the assumption concerning the outcome generation is approximately the one underlying the index.

For example, the standardized Banzhaf index indicates, for a given decision rule and resource (vote) distribution, the expected influence an actor has on the outcomes if every coalition is equally likely to form. It is often the case that all coalitions are not equally likely, for example for ideological reasons. Thus the real influence of an actor over the outcomes may grossly deviate from what one would expect on the basis of the Banzhaf index value.

Two indices, namely Shapley−Shubik and (non-standardized) Banzhaf, have been given a probabilistic interpretation by Straffin (1988). When constructing the interpretation Straffin begins with the question that we have just dealt with, to wit, what is the probability that an actor's vote will affect the decision outcome, for example that a proposal will pass if the actor supports it, but will be rejected if he opposes it?

Suppose that there are n voters and that each voter i supports a proposal with probability p_i. Straffin considers two joint probability distributions, that is, assumptions underlying the construction of vector $(p_1, \ldots, p_n)$ of individual probabilities:

1. The independence assumption: each p_i is drawn from the uniform distribution $[0, 1]$ independently of the others.
2. The homogeneity assumption: a value p is drawn from the uniform distribution $[0, 1]$ and p_i is set equal to this p, for all voters.

In the former model one assumes that each voter acts independently of the others. Moreover, since the issues to be decided are not known, any value between 0 and 1 is equally likely to describe his probability of supporting the proposal. In the latter model, each voter has an identical probability of supporting the proposal. This probability value that characterizes the entire body is chosen randomly from the real number interval between 0 and 1 so that any value has the same probability of being chosen.

Straffin shows that if the voting body satisfies the independence assumption, then a voter's (non-standardized or absolute) Banzhaf index value gives an answer to the question of his influence on the outcome. He also shows that if the voting body satisfies the homogeneity assumption, then the answer to the question we are dealing with is his Shapley–Shubik index value. In other words, if the voting body consists of independent voters, then the absolute Banzhaf index expresses the probability that the voter is decisive in the sense that if he supports the proposal, it will pass, while if he doesn't, it won't pass. His Shapley–Shubik index value, on the other hand, measures his influence on the outcome if the voting body is homogeneous.

In practice neither of these assumptions holds. Thus each one of them gives a biased picture of the 'real' power distribution. On the other hand, the Banzhaf and Shapley–Shubik indices do not often differ very much. Hence, most of the time it makes very little difference which one of them one is utilizing in measuring the *a priori* voting power distribution. In particular, in single-chamber legislatures or similar decision-making bodies using larger than simple majority rules, the indices give fairly similar results.

This is, however, not necessarily the case in multi-chamber bodies. Straffin gives the following example of a two-chamber simple majority body (Straffin 1988). In order to pass, a proposal has to be accepted by a lower house L and an upper house U. These can be considered as simple games: $L = (2; 1, 1, 1, 1)$ and $U = (3; 2, 1, 1, 1)$, where the first number in parentheses indicates the decision rule and the numbers after the semicolon indicate the vote distribution over the players. We denote the players in L by A. Since they all have the same number of votes, we do not need other symbols. In U the player with two votes is called B and the other voters of U are called C.

The power index computations become somewhat more complicated in multi-chamber bodies where the proposal has to be accepted by some majority in all houses. Just to give an example, consider the swings of A. There are three ways of choosing A's partner to form the required minimum majority in L. So, for each A there are three swings in L. For each of these the required majority in U may be formed in one of the following ways:

- B with one of C's (3 different coalitions).
- B with two of C's (3 different coalitions).
- B with all the others (1 coalition).
- All C's together (1 coalition).

So, there are eight different ways of forming a winning coalition in U. Since the proposal needs to be accepted in both houses, each A is crucial (has a swing) $3 \times 8 = 24$ times. In a similar way one computes the number of swings of each player and the weights needed for the calculation of the Shapley–Shubik index. The results are following (Straffin 1988, 73):

$$\phi_A = 0.093, \phi_B = 0.314, \phi_C = 0.105,$$

$$\beta'_A = 0.188, \beta'_B = 0.516, \beta'_C = 0.172.$$

Both indices give B the highest value, but the ranking of A and C is different. The differences between these two types of indices are, however, relatively minor when compared with those that each of them has *vis-à-vis* the Holler index.

The main difference can be expressed in terms of two monotonicity properties discussed by Allingham (1975) and Turnovec (1994).

Definition 7.3 *Let players A and B belong to a voting body which uses a fixed majority (simple or otherwise) rule in making decisions. Let the number of A's votes be strictly larger than the number of B's votes. Then an index is locally monotonic if it always assigns to A at least as large a value as to B.*

In other words, players with more resources get at least as high power index values as players with fewer resources. The locality here refers to the fact that the comparisons of values take place between players belonging to the same voting body.

Definition 7.4 *Let U and L be two voting bodies consisting of the same players so that player 1's number of votes increases from U to L, while the number of votes of all other members either decreases or remains the same from U to L. An index is globally monotonic if it gives player 1 at least as high a power index value in L as in U.*

Global monotonicity is a requirement that the power index respond to resource redistribution in a 'natural' way, that is, by not diminishing the index value of a player whose resources have increased *vis-à-vis* those of the others.

In a recent paper Turnovec shows that the Shapley—Shubik index satisfies both monotonicity conditions, while the Banzhaf index satisfies local monotonicity only. Holler's index, in turn, satisfies neither type of monotonicity (Holler and Packel 1983). Thus Holler's index may assign a player with fewer votes a value that is higher than that of a player with more votes. This is also possible when the Deegan-Packel index is used (Nurmi et al. 1997).

These properties should be kept in mind when, in the following, we apply the indices to EU institutions.

7.1.3   *A priori* voting power in the EU Council of Ministers

Since the Council of Ministers of the EU (Council, for brevity) resorts to weighted voting in many — although by no means in all — issues to be decided, it has been a natural field for applying power indices. Usually the aim of these studies is to find out the implications of various voting weight reallocations, decision rule changes or membership enlargements to *a priori* voting power distribution (Brams and Affuso 1985; Turnovec 1996). Table 7.2 shows the distribution of votes and *a priori* voting power in the current Council of Ministers of the EU assuming that the decision rule is 62 out of 87 votes. The qualified majority rule, used in the Council whenever unanimous support is not required, has always been around 70 per cent of the sum of voting weights. This is also the case today.

Since in all national parliaments the bulk of decision making takes place in the simple majority context, one possible scenario for eventual institutional reform within the EU is that the simple majority rule will be applied in the Council. What this would imply in terms of *a priori* voting power distribution can be seen in Table 7.3.

In general, and in accordance with expectations, the change would marginally benefit the largest member states at the cost of mainly the middle-sized and small countries. It is noteworthy that the Holler index value for Spain becomes larger than that of ten-vote countries, a demonstration of the local non-monotonicity of the index.

If the member countries are interested in maximizing their influence on the outcomes of Council decision making, they could aim at those decision rules for which their *a priori* voting power index values are maximal. Table 7.4 reports those decision rules for various coun-

Country	votes	ϕ	β	DP	H
France	10	0.117	0.112	0.082	0.081
Germany	10	0.117	0.112	0.082	0.0819
Italy	10	0.117	0.112	0.082	0.081
UK	10	0.117	0.112	0.082	0.081
Spain	8	0.096	0.092	0.075	0.074
Belgium	5	0.055	0.059	0.065	0.065
Greece	5	0.055	0.059	0.065	0.065
Holland	5	0.055	0.059	0.065	0.065
Portugal	5	0.055	0.059	0.065	0.065
Austria	4	0.045	0.048	0.061	0.061
Sweden	4	0.045	0.048	0.061	0.061
Denmark	3	0.035	0.036	0.057	0.058
Finland	3	0.035	0.036	0.057	0.058
Ireland	3	0.035	0.036	0.057	0.058
Luxembourg	2	0.021	0.023	0.044	0.045

Table 7.2: Power Index Values in the Council for Decision Rule 62. $\phi =$ Shapley–Shubik Index, β =Standardized Banzhaf Index, DP =Deegan-Packel Index, H =Holler Index.

Country	ϕ	β	DP	H
10-vote country	0.118	0.117	0.073	0.070
8-vote country	0.092	0.091	0.073	0.071
5-vote country	0.056	0.056	0.068	0.068
4-vote country	0.046	0.047	0.065	0.067
3-vote country	0.033	0.033	0.062	0.064
2-vote country	0.022	0.022	0.049	0.051

Table 7.3: Power Index Values in the Council for Decision Rule 44. 10-Vote Countries: France, Germany, Italy, UK; 8-Vote Country: Spain; 5-Vote Countries: Belgium, Greece, Holland, Portugal; 4-Vote Countries: Austria, Sweden; 3-Vote Countries: Denmark, Finland, Ireland; 2-Vote Country: Luxembourg.

Votes	ϕ	β	DP	H
10	78	45	67	71
	0.149	0.118	0.083	0.082
8	80	60	77	77
	0.127	0.093	0.079	0.078
5	83	83	80	80
	0.095	0.075	0.073	0.073
4	84	84	84	84
	0.086	0.075	0.071	0.071
3	85	85	85	85
	0.071	0.071	0.071	0.071
2	86	86	86	86
	0.067	0.067	0.067	0.067

Table 7.4: Optimal Decision Rules and Power Index Maxima

tries, assuming current vote distribution, at which the maximum value is reached starting from the simple majority rule.

As one would expect, the voting power maximizing decision rules are the larger, the fewer votes the countries have in the Council. The standardized Banzhaf index maximum of ten-vote states occurs at a very small value. Also the Banzhaf maximum of the eight-vote state (Spain) is achieved at a considerably smaller qualified majority than other indices would suggest. These two observations corroborate the suspicion that the critical points of the standardized Banzhaf index function do not coincide with the intuitive power maxima.

In the institutional reform debate the representatives of populous states have suggested various 'double majority' decision rules, that is, rules that would take into account not only the number of votes in the Council but also the populations of the states supporting or opposing the decisions. In particular, it has been suggested that, in addition to being supported by a qualified majority of votes in the Council, the supporting Council members ought to represent at least 50 per cent of the EU's population. The motivation for this suggestion is provided by the fact that the Council vote distribution does not exactly reflect the populations of the member states, but the least populous states are somewhat over-represented.

The effects of various double majority rules on the distribution of the voting power of various EU member states can be analysed with the aid of power indices. Table 7.5 indicates the power index value distribution in the current Council under the assumption that the decision rule 62 is

Country	Popula-tion (m)	ϕ	β	DP	H
Germany	81.2	0.117	0.112	0.082	0.081
UK	57.8	0.117	0.112	0.082	0.081
France	57.7	0.117	0.112	0.082	0.081
Italy	56.1	0.117	0.112	0.082	0.081
Spain	39.1	0.096	0.092	0.075	0.074
Holland	15.3	0.055	0.059	0.065	0.065
Greece	10.4	0.055	0.059	0.065	0.065
Belgium	10.0	0.055	0.059	0.065	0.065
Portugal	9.9	0.055	0.059	0.065	0.065
Sweden	8.7	0.045	0.048	0.061	0.061
Austria	8.0	0.045	0.048	0.061	0.061
Denmark	5.2	0.035	0.036	0.057	0.058
Finland	5.1	0.035	0.036	0.057	0.058
Ireland	3.6	0.035	0.036	0.057	0.058
Luxembourg	0.4	0.021	0.023	0.044	0.045

Table 7.5: The Double Majority Power Index Values in the Council

combined with the requirement that in order to pass a proposal in the Council the vote threshold as well as the 50 per cent population threshold has to be passed. We observe that there is practically no change from Table 7.2. The explanation is not difficult to find, to wit, any majority of countries that possess together more than 62 votes in the Council has at the same time a combined population size of at least 50 per cent of the total EU population. Thus this type of double majority changes nothing in terms of voting power. The situation is different if the simple majority principle (44 out of 87 votes) is combined the 50 per cent of the population threshold, as can be observed by comparing Table 7.2 with Table 7.6.

Since the assumptions underlying various power indices differ and give at best a very simplified picture of the decision-making situations in reality, the precise values of the indices are not of paramount interest in institutional design. Of more importance are, rather, the changes in voting power distributions that accompany various changes in the institutional framework. Double majority rules result in a significant redistribution of voting power only if the Council resorts to a simple majority to start with.

Even though the Council is at the moment the centre of decision-making power in the EU, the institutional reform debate suggests that in the future a more intensive interaction between the Council and ei-

Country	popu-lation	ϕ	β	DP	H
Germany	81.2	0.179	0.165	0.100	0.100
UK	57.8	0.139	0.134	0.081	0.079
France	57.7	0.139	0.134	0.081	0.079
Italy	56.1	0.137	0.133	0.080	0.078
Spain	39.1	0.105	0.104	0.078	0.076
Holland	15.3	0.044	0.048	0.067	0.068
Greece	10.4	0.039	0.043	0.061	0.062
Belgium	10.0	0.039	0.042	0.061	0.062
Portugal	9.9	0.039	0.042	0.061	0.062
Sweden	8.7	0.033	0.036	0.061	0.062
Austria	8.0	0.033	0.036	0.061	0.062
Denmark	5.2	0.022	0.024	0.056	0.058
Finland	5.1	0.0226	0.024	0.056	0.058
Ireland	3.6	0.020	0.023	0.055	0.057
Luxembourg	0.4	0.011	0.013	0.039	0.040

Table 7.6: The Power Index Values in the Current Council for Double Simple Majority Rule

ther the European Parliament (EP) or the Commission or both will be a plausible alternative, especially if what is called the deepening of integration (despite the probable enlargements) continues. Power indices allow us to see what those institutional interactions would imply for power distribution between various actors.

7.1.4   Voting power in a two-chamber EU

We approach the Council—Parliament interaction from the power index perspective assuming that a more direct role than is currently the case is assigned to Parliament. More specifically, let us look at the *a priori* voting power distribution in a hypothetical two-chamber system where the Parliament is the first and the Council the second chamber.

It is well known that the members of the EP do not organize themselves in national groups, but join union-wide political groupings. The membership in these groups is less permanent than in national parliaments, with EP members quite often leaving one group to join another. We shall consider each party group as a player in the two-chamber voting game. In the following calculations the data describe the EP just preceding the last union-wide elections in 1994. The abbreviations used are: PES = Party of the European Socialists (215 seats), EPP = European

Country or group	ϕ	β	DP	H
10-vote country	0.059	0.070	0.047	0.046
8-vote country	0.046	0.055	0.047	0.046
5-vote country	0.028	0.034	0.044	0.044
4-vote country	0.023	0.028	0.043	0.043
3-vote country	0.016	0.020	0.041	0.042
2-vote country	0.011	0.013	0.032	0.033
PES	0.176	0.134	0.052	0.049
EPP	0.114	0.081	0.041	0.041
UFE	0.062	0.054	0.046	0.045
ELDR	0.038	0.034	0.045	0.045
Greens	0.021	0.020	0.040	0.041
ERA	0.014	0.013	0.032	0.033
EUL/NGL	0.029	0.027	0.045	0.045
IND	0.045	0.040	0.049	0.049

Table 7.7: The Power Index Values in a Hypothetical Two-Chamber Simple Majority System

People's Party (182 seats), UFE = Union for Europe (57 seats), ELDR = European Liberal, Democratic and Reform Party (43 seats), EUL/NGL = European United Left/Nordic Green Left (33 seats), Greens = The European Green Party (29 seats), ERA = European Radical Alliance (20 seats) and IND = Non-Attached (49 seats).[3]

Table 7.7 reports the power index value distribution in a hypothetical two-chamber system in which a simple majority of votes in both chambers in required for a proposal to pass.

The big party groups are clearly the most powerful actors in this two-chamber system. Apart from that observation the entries of the table are not very informative. However, when one compares them with other hypothetical decision rules, the picture that emerges gives some suggestions as to what various modifications might entail in terms of power distribution. Table 7.8 is based on the assumption that the EP resorts to the simple majority while the Council uses the 62 decision rule.

The qualified majority rule in the Council seems to increase the voting power of the big countries dramatically *vis-à-vis* the big EP parties. One could, of course, argue that an arrangement in which one chamber

---

[3] The non-attached voters are not a party grouping on a par with the other groups. Hence its power index values are not comparable with those of the latter groups, either.

Country or group	ϕ	β	DP	H
10-vote country	0.093	0.087	0.058	0.058
8-vote country	0.076	0.072	0.053	0.053
5-vote country	0.043	0.046	0.046	0.046
4-vote country	0.035	0.037	0.044	0.044
3-vote country	0.028	0.028	0.041	0.042
2-vote country	0.016	0.018	0.032	0.032
PES	0.078	0.075	0.042	0.041
EPP	0.053	0.045	0.033	0.034
UFE	0.025	0.030	0.038	0.037
ELDR	0.015	0.019	0.037	0.037
Greens	0.008	0.011	0.033	0.034
ERA	0.005	0.008	0.026	0.027
EUL/NGL	0.011	0.015	0.037	0.037
IND	0.018	0.022	0.040	0.041

Table 7.8: The Power Index Values in a Hypothetical Two-Chamber Asymmetric Qualified Majority System

is resorting to the simple majority while the other uses a qualified one, is not symmetric. Thus, if larger than simple majorities are used, they ought to be used in both houses. Accordingly, also the EP ought to resort to a decision rule which requires about 70 per cent of votes for the passage of a motion. Table 7.9 gives the power index distribution in a symmetric qualified majority system, that is, a system in which the decision rule of the Council is 62 out of 87, while that of the EP is 417 out of 626.

In the symmetric qualified majority system the big parties of the EP would be very powerful indeed, in terms of the power indices. However, since the qualified majority system in both chambers would make any change in *status quo* rather difficult and since the simple majority rule is being used as the main legislative rule in most national parliaments, it is not likely that the symmetric qualified majority system would be adopted.

The power indices provide an instrument to trace the consequences in terms of voting power distribution ensuing from various reform proposals. This approach has been strongly criticized by many authors on several grounds. Two perhaps most often presented arguments against this approach are:

- that it ignores important considerations pertaining to coalition for-

Country or group	ϕ	β	DP	H
10-vote country	0.063	0.083	0.060	0.059
8-vote country	0.051	0.069	0.055	0.054
5-vote country	0.029	0.044	0.048	0.047
4-vote country	0.024	0.036	0.045	0.045
3-vote country	0.019	0.027	0.042	0.042
2-vote country	0.011	0.017	0.033	0.033
PES	0.272	0.115	0.073	0.072
EPP	0.117	0.104	0.058	0.054
UFE	0.018	0.007	0.025	0.027
ELDR	0.018	0.007	0.025	0.027
Greens	0.003	0.004	0.017	0.018
ERA	0.003	0.004	0.017	0.018
EUL/NGL	0.018	0.007	0.025	0.027
IND	0.018	0.007	0.025	0.027

Table 7.9: The Power Index Values in a Hypothetical Two-Chamber Symmetric Qualified Majority System

   mation (Garret and Tsebelis 1996), and
 • that it confuses power with luck (Barry 1991).

Both of these counterarguments may be correct if what one is aiming at in using power indices is a measure of real influence that the players have on concrete policy outcomes. However, in institutional design their use is of a more limited nature. What is at issue is how various institutional arrangements affect the players' structural, rule-based, possibility to change outcomes. After all, in institutional design what one is working on is the set of rules under which the political decision making takes place.

It is often noted that certain player coalitions are more likely than others. This fact can be taken into account in power index analysis by considering *a priori* unions of players. Also games that are composed of subgames — for example, consisting of interactions within a block of players — can be studied using power indices (Owen 1977; 1982; Carreras and Owen 1988). Thus, the power index analysis can be made more 'realistic' if additional information about coalition probabilities is available. It would, however, be unwise not to take advantage of the insights that other approaches provide to get a better overall picture of the workings of the EU institutions. Such approaches stem from spatial modelling and extensive-form games.

Further reading: Banzhaf (1965); Holler and Packel (1983); Deegan and Packel (1982); Johnston (1995).

7.2   SPATIAL MODELS AND EXTENSIVE-FORM GAMES

While power index analysis focuses on the distribution of voting power among members of various EU institutions, the focus of spatial modelling has thus far been on the relative power of various institutions, especially the EP and the Council. The same is true of extensive-form game modelling, which has only recently begun. What has made these latter approaches topical is their ability to shed light on the effects of various decision-making procedures.

Three difference decision-making procedures involving several EU institutions are:

- consultation procedure,
- cooperation procedure, and
- co-decision procedure.

The consultation procedure is historically the first. It is solely based on Commission—Council interaction, that is, the EP has no role in it. The cooperation procedure, introduced by the Single European Act, also involves the EP, as does the most recently introduced co-decision procedure. The power relationships between the EP and the Council have been discussed mainly in the context of the two latter procedures and the question typically asked is: does the EP have influence within EU decision making?

The cooperation procedure is the following:

1. The Commission submits a proposal to the Council.
2. Having consulted the EP (first reading), the Council may accept the proposal (with a qualified majority) or modify it (with unanimity). The result is a joint proposal.
3. The EP may accept, reject or modify the joint proposal.
4. If the EP accepts, the Council may also accept. Otherwise, the status quo prevails.
5. If the EP rejects, the Council may still accept with unanimity. Otherwise the status quo prevails.
6. If the EP modifies the joint proposal, the Commission decides whether it accepts the modification. If it accepts the modification, the Council may accept the modified proposal with a qualified majority or

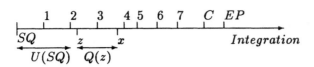

Figure 7.1: Simplified Version of Tsebelis's Model

modify it with unanimity. If the Commission rejects the EP's modification, the Council may accept the unmodified joint proposal with qualified majority or the EP's modification with unanimity.

Tsebelis (1994) discusses this procedure in the light of a spatial model. A simplified version of this model is presented in Figure 7.1 (Moser 1996; Tsebelis 1996).

Several simplifications are made for the sake of illustration. One of them is that the number of Council members is assumed to be seven. Moreover, there is only one policy dimension, the attitude towards deepending of integration. It is usually assumed that not only the Commission but also the EP support integration to a larger extent than the Council. In Figure 7.1 the bliss points of the Council members are represented by numbers $1, \ldots, 7$ and those of the Council and EP by C and EP, respectively. The status quo is indicated by SQ. $U(SQ)$ denotes the policies that are unanimously preferred to the status quo, while $Q(z)$ denotes those policies that are supported by a qualified majority (assumed to be 4 out of 7 members of the Council) to policy z. In the model both the EP and the Commission are viewed as unitary actors, which is another simplification.

Tsebelis argues that the cooperation procedure gives the EP some conditional agenda-setting power. This can be seen in the second reading phase of the EP where it can modify the joint proposal of the Council and the Commission. The modification can be made in such a way that accepting it is easier for the Council than modifying it. This is illustrated in Figure 7.1. To get a conservative estimate of the conditional agenda-setting power of the EP, we assume that the joint proposal is as much to the right along the integration dimension as possible. This point is z since it is unanimously preferred to the status quo by the Council. In Figure 7.1 z is as close as SQ to the ideal point of the least prointegration member 1. In other words, 1 is indifferent between z and SQ. Now, making a modification x, the EP can get as near as possible to its ideal point and yet get the modification passed in the Council. This follows from the fact that four members of the Council consider x better than z or SQ. The distance between x and z gives now a conservative estimate of EP's conditional agenda-setting power.

Tsebelis's model has been criticized in the literature (Moser 1996). For our purposes it is, however, a suitable illustration of spatial modelling for institutional design. One could argue that a one-dimensional representation is not adequate to model real-life decision-making situations. In fact Tsebelis's model is two-dimensional, but the crucial point it makes can be discussed using the one-dimensional projection. The conclusion that Tsebelis draws from his model is that the EP has some real decision-making power in the cooperation procedure. This conclusion is of a qualitative nature, which is an advantage, since when several dimensions are introduced one needs to introduce a norm, that is, a distance measure, as well. This, in turn, may well differ from one decision maker to another, making the conclusions more difficult if not impossible to draw. It can, however, be argued that qualitatively speaking the EP has no power at all in the cooperation procedure. This argument can be made on the basis of extensive-form games.

As each EP decision-making procedure involves well-defined stages and decision rules to be applied in them, it seems natural to describe the process as an extensive-form game of perfect recall. At each stage of the procedure the decision-making body whose turn it is knows the outcome of the previous stages. Therefore, one could in principle resort to Zermelo's algorithm in finding out the solutions to various procedures. This is precisely what Annick Laruelle has done in a recent paper (Laruelle 1997).

The oldest EU decision-making procedure, the consultation procedure, can be represented as an extensive-form game, as in Figure 7.2.

In modelling the payoffs and strategies Laruelle adopts the spatial modelling apparatus in assuming that the decisions can be represented as real numbers in some interval on which the players have their ideal points. The Commission's ideal point is denoted by C, the Council's (now assumed to be a unitary player) by M and EP's by P. The norm is defined as follows: $U_i(x) = - \mid x - i \mid$, for $i = C, M, P$. In other words, the larger the absolute value of the difference between a player's ideal point and the point under consideration, the smaller the utility value of the latter to the player. The status quo is represented by the value $x = 0$. Once a proposal x is given, the amendment of it is a policy in some neighbourhood of x. The neighbourhood is assumed to be given exogenously so that points amendments to x are within the interval $[x - \tau, x + \tau]$, where τ is a constant.

In the consultation procedure C first makes a proposal, that is, gives a value x or else the game ends at the status quo. Second, the Council

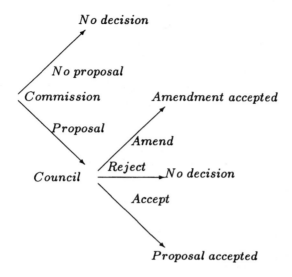

Figure 7.2: Consultation Procedure

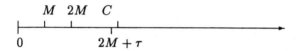

Figure 7.3: Equilibrium in Consultation Procedure

accepts x, rejects it or makes an amendment A_M to it so that $A_M \in [x - \tau, x + \tau]$. The outcome is determined directly by the Council's decision: it is $x = 0$, if it rejects the proposal, x if it accepts it and $x = A_M$ if it amends it. The outcomes ensuing from backwards induction differ, depending on the positions of the ideal points of the Commission and the Council. Let us consider one case for the sake of illustration. Assume that $C > 2M > 0$ (see Figure 7.3).

The Commission strives at an outcome as far to the right as possible, but knowing that if its proposal is too extreme, the Council will prefer 0 to it. As the Commission, on the other hand, prefers any policy to the right of 0 to 0 itself, it will in the equilibrium propose a policy that, once it has been modified by the Council within range τ, is closer to the latter's ideal point than 0 is. $2M + \tau$ is such a policy proposal by the

Commission, since when the Council modifies it by bringing it as close to its ideal point M as possible, the outcome is $2M$, which is exactly as far away from M as 0 is. Any proposal of the Commission that is further to the right will be vetoed by the Council and any proposal left of $2M + \tau$ will take the outcome after the Council's decision further away from the Commission's ideal point. Hence in equilibrium the Commission proposes $2M + \tau$ and the Council responds by $2M$, which will be the outcome.

Laruelle goes through all possible ideal point configurations and finds that the equilibrium positions S always satisfy the following:

$$S = \max\{0, \min\{2M, C\}\}.$$

This seems plausible in the consultation procedure. Considerably more surprising is her finding in the cooperation procedure described. To wit, she shows that the equilibrium outcomes satisfy precisely the same condition as the one just stated for S. In other words, despite the fact that the EP seems to play a role in the cooperation procedure, its preferences do not appear in the definition of equilibrium outcomes. This seems to contradict the conclusions of Tsebelis.

The third EU decision-making procedure, the co-decision procedure, was introduced in the Maastricht Treaty. It appears to give the EP more influence on the decision outcomes than the previous ones. The co-decision procedure is identical with the cooperation procedure up to the point when the EP enters the game. If it rejects the proposal previously proposed by the Commission and subsequently accepted by the Council, the proposal will be modified or rejected by a special conciliation committee. This committee consists of the same number of representatives of both the Council and the EP. If the EP amends the proposal it goes back to the Commission as in the cooperation procedure. If the latter rejects it, the proposal will once more be dealt with by the Council (as in the cooperation procedure), but its rejection will not necessarily kill the amended proposal as in the cooperation procedure, but the conciliation committee will have the final say. If the Commission accepts the EP's amendment, the proposal goes again to the Council, which may finally accept the amended version. If the Council rejects it, the issue will be handed to the conciliation committee. The novelty of the system is thus the conciliation committee, which has to a degree assumed the previous tasks of the Council.

Laruelle shows that the co-decision procedure has somewhat improved the EP's influence on the equilibrium outcomes. The subgame-perfect equilibrium policies S, derived by backwards induction, satisfy the following conditions:

$$S = \max\{0, \min\{2M, 2M/P, C\}\},$$

where M/P is the position of the conciliation committee. This, in turn, is assumed to be the average of the Council's and the EP's positions. Since the latter assumption is of a purely *ad hoc* nature, it is difficult to say whether the improvement in the EP's position in subgame-perfect equilibrium outcomes is essentially better than in the previous procedures.

The institutions of the EU have turned out to be a challenging field of application of the methods of institutional design. The main contribution of those methods is in tracing the likely consequences of various new arrangements to the parties involved in the design process. At best the methods can assist in predicting whether proposed institutions are likely to have their intended effects.

7.3 BIBLIOGRAPHICAL REMARKS

The main focus of the research on the institutional framework of the EU has been on the power of the EP. The books by Jacobs, Corbett and Westlake as well as the volume edited by Keohane and Hoffman provide overviews of the problematique (Jacobs and Corbett 1990; Westlake 1994; Keohane and Hoffman 1991). The Council has been an obvious field for application of power indices for a long time. The earliest studies were carried out by Brams and Affuso some twenty years ago. In the 1990s a new wave of power index studies has emerged in the literature inspired by the various enlargements of the community as well as by the calls for greater popular influence over the outcomes of EU decision making. A reasonably representative sample of this more recent work consists of articles by Brams and Affuso (1985), Brückner and Peters (1996), Felsenthal and Machover (1997), Lane et al. (1996), Turnovec (1996) and Widgrén (1994).

The power indices are introduced in several texts. Brams's (1975) is a relatively non-technical account of the best-known indices. For a reader who plans to apply power indices in compound games, that is, games consisting of subgames, it is absolutely necessary to consult Owen's (1982) book as well as the collections edited by Holler (1982b) and Roth (1988). The occasional discrepancies between power indices raise the question of whether a meaningful single-valued index can ever be found. The recent article by Taylor and Zwicker (1997) takes a new approach and discusses interval-valued power measures. Berg (1997) gives a very illuminating discussion on the relationships between the Shapley–Shubik, Banzhaf and Deegan-Packel indices.

The power index approach to EU institutions has been criticized by Garrett and Tsebelis (1996; Tsebelis and Garrett 1996). These authors favour spatial modelling since it makes it possible to analyse the coalition possibilities taking into account the players' positions *vis-à-vis* each other on policy dimensions. The spatial models have also been applied by Hubschmid and Moser (1997), Moser (1997) and Steunenberg (1994). Some recent works combine the spatial and extensive form game modelling (Laruelle 1997; Laruelle and Widgrén 1997; Steunenberg 1996).

8 Conclusion

Two themes run through the history of political thought: how to educate great men and how to make wise laws. In late medieval and early modern times much work was devoted to advising rulers in methods that would bring glory to them and their commonwealths. Since the human nature was commonly thought to be corruptible, the most important task was to issue laws and establish institutions that would counteract the inevitable slide towards corruption and chaos, in other words, towards the rule of Fortuna.

In our times the prevalent view of institutions is not essentially different. Societies are more or less defined by institutions; they affect practically all aspects of human life. The goals that institutions are thought to have are, however, less elevated or abstract than those that the medieval thinkers had in mind. Institutions operate and produce outcomes. Thus it is natural to evaluate them in the light of those outcomes. It is especially worthwhile to assess them with regard to the extent that the goals they are intended to serve are, in fact, achieved. Evaluations of this type are the content of much, if not most, political debate. However, not only are existing institutions being evaluated, but also new ones are being designed and old ones modified.

The theoretical approaches dealt with in the preceding provide tools for evaluating institutions as well as for designing them. The tools we have discussed stem from the rational actor model. The settings in which this model yields its strongest results are very simplified representations of societies. It is, therefore, unlikely that these tools by themselves would be sufficient for enlightened institutional design. Much more is needed: for example, comparative research on institutional performance in various cultural settings, an understanding of patterns of cultural and political socialization, the process of creation and transformation of traditions, the emergence of values (such as justice) and so on. These complementary studies are needed to provide new insights into the operation of institutions, to suggest hypotheses and to test constructed models.

Over time the rational actor models have grown more and more complex. New factors or variables that were previouly held exogenous or 'given', are endogenized, that is, regarded as parts of the model. Thus the rational actor models of institutions are becoming more comprehensive all the time. Indeed, the models themselves often become the main object of study. Models are solved, simulations are conducted on them and parameters are adjusted for stability. In some cases straightforward experiments are performed to test hypotheses in moderately complex individual choice or game situations. These are steps towards widening the scope of rational actor models. The strategy of model development ranges from small and fairly well understood — that is, rational actors in well-defined choice situations — to large and less well-known — that is, institutions, organizations and societies.

Although the rational actor model has been the sole focus in the preceding, it makes no sense to argue that it is the only legitimate approach to institutional design. As was just pointed out, the comparative study of institutions in various types of settings complements the picture that one is able to outline on the basis of the rational actor model. But the benefit is mutual: comparative research can also derive testable hypotheses from rational actor models and use the model to help in understanding why institutions perform in the way they do in various circumstances.

Most institutions in the world are constructed on the basis of intuition, past experience or simply hunch. Since many institutions have a very long history (for example in the case of many constitutions, party systems or property laws), it is likely that one can learn a great deal by systematic study of them. A suitable systematic background for this study is provided by the theories of rational behaviour. Thus the theories dealt with in this book can also be applied in comparative work.

Bibliography

Abrams, R. (1980), *Foundations of Political Analysis*, New York: Columbia University Press.

Abreu, D. and A. Sen (1990), 'Subgame perfect implementation: a necessary and almost sufficient condition', *Journal of Economic Theory*, **50**, 285–99.

Achinstein, P. (1968), *Concepts of Science: A Philosophical Analysis*, Baltimore: The Johns Hopkins University Press.

Aizerman, M. and F. Aleskerov (1995), *Theory of Choice*, Amsterdam: North-Holland.

Allais, M. (1979), 'The foundations of a positive theory of choice involving risk and a criticism of the postulates and axioms of the American school', in M. Allais and O. Hagen (eds), *The Expected Utility Hypothesis and the Allais Paradox*, Dordrecht: D. Reidel.

Allais, M. and O. Hagen (eds), *The Expected Utility Hypothesis and the Allais Paradox*, Dordrecht: D. Reidel.

Allingham, M.G. (1975), 'Economic power and values of games', *Zeitschrift für Nationalökonomie*, **35**, 293–9.

Anscombe, F.J. and R.J. Aumann (1963), 'A Definition of Subjective Probability', *Annals of Mathematical Statistics*, **34**, 199–205.

Anscombe, G.E.M. (1976), 'On frustration of the majority by fulfilment of the majority's will', *Analysis*, **36**, 161–8.

Arrow, K.J. (1963), *Social Choice and Individual Values*, 2nd ed., New York: Wiley.

Arrow, K.J. (1986), 'Agency and the market', in K.J. Arrow and M.D. Intriligator (eds), *Handbook of Mathematical Economics*, **III**, Amsterdam: North-Holland.

Axelrod, R. (1980a), 'Effective choice in the prisoner's dilemma', *Journal of Conflict Resolution*, **24**, 3–25.

Axelrod, R. (1980b), 'More effective choice in the prisoner's dilemma', *Journal of Conflict Resolution*, **24**, 379–403.

Axelrod, R. (1981),'The emergence of cooperation among egoists', *The American Political Science Review*, **75**, 306–18.

Axelrod, R. (1984a), *The Evolution of Cooperation*, New York: Basic Books.

Axelrod, R. (1984b), 'The problem of cooperation', in T. Cowen (ed.), *The Theory of Market Failure: A Critical Examination*, Fairfax, Va.: George Mason University Press.

Axelrod, R. and D. Dion (1988), 'The further evolution of cooperation', *Science*, **242**, 1385–90.

Axelrod, R. and W.D.Hamilton (1981), 'The evolution of cooperation', *Science*, **211**, 1390-96.

Baird, D.G., R.H. Gertner and R.C. Picker (1994), *Game Theory and the Law*, Cambridge, MA: Harvard University Press.

Banks, J.S. (1985), 'Sophisticated voting outcomes and agenda control', *Social Choice and Welfare*, **4**, 295–306.

Banks, J.S. (1989), 'Equilibrium outcomes in two-stage amendment procedures', *American Journal of Political Science*, **33**, 25–43.

Banks, J.S. (1995), 'Singularity theory and core existence in the spatial model', *Journal of Mathematical Economics*, **24**, 523–36.

Banzhaf, J.F, III (1965), 'Weighted voting doesn't work: a mathematical analysis', *Rutgers Law Review*,**19**, 317–43.

Bar-Hillel, M. and A. Margalit (1988), 'How vicious are cycles of intransitive choice?', *Theory and Decision*, **24**, 119–45.

Barry, B. (1991), 'Is it better to be powerful than lucky?', in B. Barry, *Democracy and Power*, Oxford: Clarendon Press.

Bartholdi, J.J., C.A. Butler and M.A. Trick (1986), 'More on the evolution of cooperation', *Journal of Conflict Resolution*, **30**, 129–40.

Bartholdi, J.J. and J.B. Orlin (1991), 'Single transferable vote resists strategic voting', *Social Choice and Welfare*, **8**, 341–54.

Bartholdi, J.J., C.A. Tovey and M.A. Trick (1989), 'The computational difficulty of manipulating an election', *Social Choice and Welfare*, **6**, 227–41.

Bell, D.E. (1982), 'Regret in decision making under uncertainty', *Operations Research*, **30**, 961–81.

Berg, S. (1996), 'Condorcet's jury theorem and the reliability of majority voting', *Group Decision and Negotiation*, **5**, 229–38.

Berg, S. (1997), 'On voting power indices and a class of probability distributions', *Group Decision and Negotiation*, forthcoming.

Bezembinder, Th. and P. Van Acker (1985), 'The Ostrogorski paradox and its relation to nontransitive choice', *Journal of Mathematical Sociology*, **11**, 131–58.

Bianco, W.T., P.C.Ordeshook and G. Tsebelis (1990), 'Crime and punishment: are one-shot games enough?', *American Political Science Review*, **84**, 569–86.

Binmore, K. (1987), 'Modeling rational players. Part I', *Economics and Philosophy*, **3**, 179–214.

Binmore, K. (1988), 'Modeling rational players. Part II', *Economics and Philosophy*, **4**, 9–55.

Binmore, K. (1992), *Fun and Games*, Lexington, MA: D.C. Heath and Company.

Black, D. (1958), *Theory of Committees and Elections*, Cambridge: Cambridge University Press.

Blair, D.H. (1981), 'On the ubiquity of strategic voting opportunities', *International Economic Review*, **22**, 649–55.

Brams, S.J. (1975), *Game Theory and Politics*, New York: Free Press.

Brams, S.J. (1977), 'Deception in 2×2 Games', *Journal of Peace Science*, **2**, 171–203.

Brams, S.J. (1980), *Biblical Games*, Cambridge, MA: MIT Press.

Brams, S.J. (1983), *Superior Beings: If They Exist, How Would We Know?*, New York: Springer.

Brams, S.J. (1985), *Rational Politics*, Washington, DC: CQ Press.

Brams, S.J. (1994), *Theory of Moves*, Cambridge: Cambridge University Press.

Brams, S.J. and P.J. Affuso (1985), 'New paradoxes of voting power in the EC council of ministers', *Electoral Studies*, **4**, 135–9.

Brams, S.J. and P.C. Fishburn (1983), *Approval Voting*, Boston: Birkhäuser.

Brams, S.J., D.M. Kilgour and W.S. Zwicker (1993), 'A new paradox of vote aggregation', paper presented at the 1993 Annual Meeting of the American Political Science Association; mimeo, New York University, Department of Politics.

Brams, S.J., D.M. Kilgour and W.S. Zwicker (1997), 'The paradox of multiple elections', *Social Choice and Welfare*, forthcoming.

Brams, S.J. and W. Mattli (1993), 'Theory of moves: overview and examples', *Conflict Management and Peace Science*, **12**, 1–39.

Brams, S.J. and D. Wittman (1981), 'Non-myopic equilibria in 2 × 2 games', *Conflict Management and Peace Science*, **6**, 39–62.

Brandenburger, A. (1992), 'Knowledge and equilibrium in games', *Journal of Economic Perspective*, **6**, 83–101.

Brückner, M. and T. Peters (1996), 'Further evidence on EU voting power', *Journal of Theoretical Politics*, **8**, 415–19.

Buchanan, J.M. (1975), *The Limits of Liberty*, Chicago: The University of Chicago Press.

Buchanan, J.M. and G. Tullock (1962), *Calculus of Consent*, Ann Arbor: University of Michigan Press.

Calvert, R.C. (1987), 'Reputation and legislative leadership', *Public Choice*, **55**, 81–119.

Carreras, F. and G. Owen (1988), 'Evaluation of the Catalonian parliament 1980–1984', *Mathematical Social Sciences*, **15**, 87–92.

Chamberlin, J.R. (1985), 'An investigation into the relative manipulability of four voting systems', *Behavioral Science*, **30**, 195–203.

Chernoff, H. (1954), 'Rational selection of decision functions', *Econometrica*, **22**, 422–43.

Chernoff, H. and L.E. Moses (1957), *Elementary Decision Theory*, New York: Wiley.

Coleman, J.S. (1990), *Foundations of Social Theory*, Cambridge, MA: Harvard University Press.

Colman, A. and I. Pountney (1978), 'Borda's voting paradox: theoretical likelihood and electoral occurrences', *Behavioral Science*, **23**, 15–20.

Colomer, J. M. (1991), 'Transitions by agreement: modeling the Spanish way', *The American Political Science Review*, **85**, 1283–302.

Coughlin, P.J. (1992), *Probabilistic Voting Theory*, Cambridge: Cambridge University Press.

Cowan, Th. A. and P.C. Fishburn (1988), 'Foundations of preference', in G. Eberlein and H. Berghel (eds), *Theory and Decision*, Dordrecht: D. Reidel.

Danilov, V. (1992), 'Implemenation via Nash-equilibria' *Econometrica*, **60**, 43–56.

Daudt, H. and D. Rae (1978), 'Social contract and the limits of majority rule', in P. Birnbaum, J. Lively and G. Parry (eds), *Democracy, Consensus & Social Contract*, London and Beverly Hills: SAGE Publications.

Davis, O.A., M.H. DeGroot and M.J. Hinich (1972), 'Social preference orderings and majority rule', *Econometrica*, **40**, 147–57.

Debreu, G. (1959), *Theory of Value*, New York: Wiley.

DeGrazia, A. (1953), 'Mathematical derivation of an election system', *Isis*, **44**, 42–51.

Deegan, J. and E.W. Packel (1982), 'To the (minimal winning) victors go the (equally divided) spoils: A new power index for simple *n*-person games', in S.J. Brams, W.F. Lucas and P.D. Straffin (eds), *Political and Related Models in Applied Mathematics*, New York: Springer-Verlag.

Dowding, K. (1997), 'Equity and voting: Why democracy needs dictators', *L'Annee sociologique*, **47**, 39–53.

Downs, A. (1957), *An Economic Theory of Democracy*, New York: Harper & Row.

Dubey, P. and L.S. Shapley (1979), 'Mathematical properties of the Banzhaf power index', *Mathematics of Operations Research,* **4**, 99–130.

Dummett, M. (1984), *Voting Procedures,* Oxford: Clarendon Press.

Dummett, M. and R. Farquharson (1961), 'Stability in voting', *Econometrica,* **29**, 33–43.

Ellsberg, D. (1961), 'Risk, ambiguity, and the Savage axioms', *Quarterly Journal of Economics,* **75**, 643–69.

Elster, J. (1979), *Ulysses and the Sirens: Studies in Rationality and Irrationality,* Cambridge: Cambridge University Press.

Elster, J. (1996), 'The constitution-making process', in J. Casas Pardo and F. Schneider (eds), *Current Issues in Public Choice,* Cheltenham: Edward Elgar.

Enelow, J.M. and M.J. Hinich (1983), 'On Plott's pairwise symmetry condition for majority rule equilibrium', *Public Choice,* **40**, 317–21.

Enelow, J.M. and M.J. Hinich (1984), *The Spatial Theory of Voting. An Introduction,* Cambridge: Cambridge University Press.

Enelow, J.M. and M.J. Hinich (eds) (1990), *Advances in the Spatial Theory of Voting,* Cambridge: Cambridge University Press.

Farquharson, R. (1969), *Theory of Voting,* New Haven: Yale University Press.

Feldman A.M. (1980), *Welfare Economics and Social Choice Theory,* Boston: Martinus Nijhoff.

Felsenthal, D.S. and M. Machover (1997), 'The weighted voting rule in the EU's council of ministers, 1958-95: intentions and outcomes', *Electoral Studies,* **16**, 33–47.

Ferejohn, J.A. and D.M. Grether (1974), 'On a class of rational social decision procedures', *Journal of Economic Theory,* **8**, 471–82.

Fishburn, P.C. (1970), *Utility Theory for Decision Making,* New York: Wiley.

Fishburn, P.C. (1973), *The Theory of Social Choice,* Princeton: Princeton University Press.

Fishburn, P.C. (1974), 'Paradoxes of voting', *American Political Science Review,* **68**, 537–46.

Fishburn, P.C. (1977), 'Condorcet social choice functions', *SIAM Journal of Applied Mathematics,* **33**, 469–89.

Fishburn, P.C. (1988), *Nonlinear Preference and Utility Theory,* Baltimore: The Johns Hopkins University Press.

Fishburn, P.C. (1991), 'Nontransitive preferences in decision theory', *Journal of Risk and Uncertainty,* **4**, 113–34.

Fishburn, P.C. and S.J. Brams (1983), 'Paradoxes of preferential voting', *Mathematics Magazine,***56**, 201–14.

French, S. (1986), *Decision Theory: an Introduction to the Mathematics of Rationality*, Chichester: Ellis Horwood.

French, S. and Z. Xie (1994), 'A perspective on recent developments in utility theory', in S. Rios (ed.), *Decision Theory and Decision Analysis: Trends and Challenges*, Boston–Dordrecht–London: Kluwer.

Friedman, M. and L. Savage (1952), 'The expected-utility hypothesis and measurability of utility', *Journal of Political Economy*, **60**, 463–74.

Fudenberg, D. and J. Tirole (1991), *Game Theory*, Cambridge: The MIT Press.

Gärdenfors, P. (1976), 'Manipulation of social choice functions', *Journal of Economic Theory*, **13**, 217–28.

Garrett, G. and G. Tsebelis (1996), 'An institutional critique of inter-governmentalism', *International Organization*, **50**, 269–99.

Gauthier, D. (1996), 'Commitment and choice', in F. Farina, S. Vannucci and F. Hahn (eds), *Ethics, Rationality and Economic Behaviour*, Oxford: Oxford University Press.

Gauthier, D. (1997), 'Resolute choice and rational deliberation: a critique and a defense', *Nous*, **31**, 1–25.

Gibbard, A. (1973), 'Manipulation of voting schemes: a general result', *Econometrica*, **41**, 587–601.

Gibbard, A. (1977), 'Manipulation of schemes that mix voting with chance', *Econometrica*, **45**, 665–81.

Giere, R. (1979), *Understanding Scientific Reasoning*, New York: Holt, Rinehart and Winston.

Gorman, J.L. (1978), 'A problem in the justification of democracy', *Analysis*, **39**, 46–50.

Greenberg, J. (1979), 'Consistent majority rule over compact sets of alternatives', *Econometrica*, **47**, 627–36.

Grether, D.M. and C.R. Plott (1979), 'Economic theory of choice and the preference reversal phenomenon', *American Economic Review*, **69**, 623–38.

Grofman, B. and A. Lijphart (eds), *Electoral Laws and Their Political Consequences*, New York: Agathon Press.

Groves, Th. and J. Ledyard (1977), 'Optimal allocation of public goods: a solution to the "free rider" problem', *Econometrica*, **45**, 783–809.

Guttman, J.M. (1996), 'Unanimity and majority rule: an economic analysis', *European Journal of Political Economy*, forthcoming.

Hamburger, H. (1973), 'N-person prisoner's dilemma', *Journal of Mathematical Sociology*, **III**, 27–48.

Hamburger, H. (1979), *Games as Models of Social Phenomena*, San Francisco: Freeman.

Hammond, P. (1988), 'Consequentialist foundations for expected utility', *Theory and Decision*, **25**, 25–78.

Hardin, R. (1971), 'Collective action as an agreeable n-prisoner's dilemma', *Behavioral Science*, **16**, 472–81.

Hardin, R. (1982), *Collective Action*, Baltimore: The Johns Hopkins University Press.

Hardin, R. (1995), *One for All: The Logic of Group Conflict*, Princeton: Princeton University Press.

Harsanyi, J.C. (1977), *Rational Behavior and Bargaining Equilibrium in Games and Social Situations*, Cambridge: Cambridge University Press.

Harsanyi, J.C. and R. Selten (1988), *A General Theory of Equilibrium Selection in Games*, Cambridge, MA: The MIT Press.

Herne, K. (1997a), 'Decoy alternatives in policy choices: asymmetric domination and compromise effects', *European Journal of Political Economy*, **13**, 575–89.

Herne, K. (1997b), *Decoy Alternatives in Individual Choice and Politics*, Turku: Annales Universitatis Turkuensis, Ser. B-220.

Herne, K. and H. Nurmi (1993), 'The distribution of a priori voting power in the EC council of ministers and the European Parliament', *Scandinavian Political Studies*, **16**, 269–84.

Holler, M.J. (1982a), 'Forming coalitions and measuring voting power', *Political Studies*, **30**, 262–71.

Holler, M.J. (ed.) (1982b), *Power, Voting and Voting Power*, Würtzburg-Wien: Physica–Verlag.

Holler, M.J. (1993), 'Fighting pollution when decisions are strategic', *Public Choice*, **76**, 347–56.

Holler, M.J. and E.W. Packel (1983), 'Power, luck and the right index', *Zeitschrift für Nationalökonomie*, **43**, 21–9.

Howard, N. (1971), *Paradoxes of Rationality*, Cambridge, MA: MIT Press.

Huber, J., J.W. Payne and C. Puto (1982), 'Adding asymmetrically dominated alternatives: violations of regularity and similarity hypothesis', *Journal of Consumer Research*, **9**, 90–98.

Hubschmid, C. and P. Moser (1997), 'The co-operation procedure in the EU: why was the European parliament influential in the decision on car emission standards?', *Journal of Common Market Studies*, **35**, 225–42.

Hurwicz, L. (1972), 'On informationally decentralized systems', in R. Radner and C.B. McGuire (eds), *Decision and Organization: A Volume in Honor of Jacob Marschak*, Amsterdam: North-Holland.

Hurwicz, L. (1979), 'On allocations attainable through Nash equilibria', in J.-J. Laffont (ed.), *Aggregation and Revelation of Preferences*, Amsterdam: North-Holland.

Intriligator, M.D. (1971), *Mathematical Optimization and Economic Theory*, Englewood Cliffs: Prentice-Hall.

Intriligator, M.D. (1973), 'A probabilistic model of social choice', *Review of Economic Studies*, 40, 553–60.

Intriligator, M.D. (1982), 'Probabilistic models of choice', *Mathematical Social Sciences*, 2, 157–66.

Ishikawa, S. and K. Nakamura (1979), 'The strategy-proof social choice functions', *Journal of Mathematical Economics*, 6, 283–95.

Jacobs, F. and R. Corbett (1990), *The European Parliament*, Harlow: Longman.

Johnston, R.J. (1995), 'The conflict over qualified majority voting in the European Union council of ministers: an analysis of the UK negotiating stance using power indices', *British Journal of Political Science*, 25, 245–88.

Jungermann, H. (ed.) (1977), *Decision Making and Change in Human Affairs*, Dordrecht: D. Reidel.

Kahneman, D. and A. Tversky (1979), 'Prospect theory: an analysis of decision under risk', *Econometrica*, 47, 263–91.

Katz, R.S. (1980), *A Theory of Parties and Electoral Systems*, Baltimore: Johns Hopkins Press.

Keck, O. (1987), 'The information dilemma: private information as a cause of transaction failure in markets, regulation, hierarchy and politics', *Journal of Conflict Resolution*, 31, 139–63.

Keck, O. (1988), 'A theory of white elephants: asymmetric information in government support for technology', *Research Policy*, 17, 187–201.

Keck, O. (1993), *Information, Macht und gesellshaftliche Rationalität*, Baden-Baden: Nomos Verlagsgesellschaft.

Kelly, J.S. (1978), *Arrow Impossibility Theorems*, New York: Academic Press.

Kelly, J.S. (1988), *Social Choice Theory: An Introduction*, Berlin: Springer-Verlag.

Kelly, J.S. (1991), 'Social choice bibliography', *Social Choice and Welfare*, 8, 97–169.

Keohane, R.O. and S. Hoffman (eds) (1991), *The New European Community: Decision-Making and Institutional Change*, Boulder: Westview Press.

Kramer, G.H. (1973), 'On a class of equilibrium conditions for majority rule', *Econometrica*, 41, 285–97.

Kramer, G.H. (1977), 'A dynamical model of political equilibrium', *Journal of Economic Theory*, **16**, 310–34.

Krantz, D.H., D. Luce, P. Suppes and A. Tversky (1971), *Foundations of Measurement*, vol. 1, San Diego: Academic Press.

Krantz, D.H., D. Luce, P. Suppes and A. Tversky (1989), *Foundations of Measurement*, vol. 2, San Diego: Academic Press.

Krantz, D.H., D. Luce, P. Suppes and A. Tversky (1990), *Foundations of Measurement*, vol. 3, San Diego: Academic Press.

Kreps, D.M. (1990), *Game Theory and Economic Modelling*, New York: Oxford University Press.

Laakso, M. (1975), *Eduskunta koalitio- ja valtasuhderakenteena*, Helsinki: Acta Politica, Fasc. IX.

Lagerspetz, E. (1984), 'Money as a social contract', *Theory and Decision*, **17**, 1–9.

Lagerspetz, E. (1995), 'Paradoxes and representation', *Electoral Studies*, **15**, 83–92.

Lane, J.-E., R. Mæland and S. Berg (1996), 'Voting power under the EU constitution', *Journal of Theoretical Politics*, **7**, 223–30.

Laruelle, A. (1997), 'The EU decision-making procedures: some insights from non-cooperative game theory', mimeo, IRES, Department of Economics, Université catholique de Louvain.

Laruelle, A. and M. Widgrén (1997), 'The development of the division of power between EU commission, EU council and European parliament', The Research Institute of the Finnish Economy, Discussion papers, No. 584.

Ledyard, J.O. (ed.) (1995), *The Economics of Informational Decentralization: Complexity, Efficiency, and Stability. Essays in Honor of Stanlay Reiter*, Norwell, MA: Kluwer.

Leinfellner, W. (1986), 'The prisoner's dilemma and its evolutionary iteration', in A.Diekman and P. Mitter (eds), *Paradoxical Effects of Social Behavior: Essays in Honor of Anatol Rapoport*, Wien: Physica-Verlag.

Lichtenstein, S. and P. Slovic (1971), 'Reversal of preferences between bids and choices in gambling decisions', *Journal of Experimental Psychology*, **89**, 46–55.

Lijphart, A. and B. Grofman (eds) (1984), *Choosing an Electoral System*, New York: Praeger.

Loomes, G. and R. Sugden (1982), 'Regret theory: an alternative theory of rational choice under uncertainty', *The Economic Journal*, **92**, 805–24.

Loomes, G. and R. Sugden (1986), 'Disappointment and dynamic consistency in choice under uncertainty', *Review of Economic Studies*, **53**, 271–82.

Luce, R.D. (1959), *Individual Choice Behavior*, New York: Wiley.

Luce, R.D. and H. Raiffa (1957), *Games and Decisions*, New York: Wiley.

MacCrimmon, K.R. and S. Larsson (1979), 'Utility theory: axioms versus "paradoxes"', in M. Allais and O. Hagen (eds), *The Expected Utility Hypothesis and the Allais Paradox*, Dordrecht: D. Reidel.

Machina, M. (1982), 'Expected utility analysis without the independence axiom', *Econometrica*, **50**, 277–323.

MacKay, A.F. (1980), *Arrow's Theorem: The Paradox of Social Choice*, New Haven: Yale University Press.

Marschak, J. and R. Radner (1972), *Economic Theory of Teams*, New Haven: Yale University Press.

Maskin, E.S. (1985), 'The theory of implementation in Nash equilibrium', in L. Hurwicz, D. Schmeidler and H. Sonnenschein (eds), *Social Goals and Social Organization: Essays in Memory of Elisha Pazner*, Cambridge: Cambridge University Press.

May, K.O. (1952), 'A set of independent, necessary and sufficient conditions for simple majority decision', *Econometrica*, **20**, 680–84.

McClennen, E.F. (1990), *Rationality and Dynamic Choice*, Cambridge: Cambridge University Press.

McCubbins, M.D. and Th. Schwartz (1985), 'The politics of flatland', *Public Choice*, **46**, 45–60.

McKelvey, R.D. (1976), 'Intransitivities in multidimensional voting models and some implications for agenda control', *Journal of Economic Theory*, **12**, 472–82.

McKelvey, R.D. (1979), 'General conditions for global intransitivities in formal voting models', *Econometrica*, **47**, 1085-112.

McKelvey, R.D. and R.G. Niemi (1978), 'A multistage game representation of sophisticated voting for binary procedures', *Journal of Economic Theory*, **18**, 1–22.

McKelvey, R.D. and N. Schofield (1986), 'Generalized symmetry conditions at a core point', *Econometrica* **55**, 923–34.

McLean, I. (1996), 'E.J. Nanson, social choice and electoral reform', *Australian Journal of Political Science*, **31**, 369–85.

McLean, I. and A.B. Urken (eds) (1995), *Classics of Social Choice*, Ann Arbor: The University of Michigan Press.

Miller, N.R. (1977), 'Graph-theoretical approaches to the theory of voting', *American Journal of Political Science*, **21**, 769–803.

Miller, N.R. (1980), 'A new solution set for tournaments and majority voting: further graph-theoretical approaches to the theory of voting', *American Journal of Political Science*, **24**, 68–96.

Miller, N.R. (1986), 'Information, electorates, and democracy: some extensions and interpretations of the Condorcet jury theorem', in B. Grofman and G. Owen (ed.), *Information Pooling and Group Decision Making*, Greenwich, CT: JAI Press.

Miller, N.R. (1995), *Committees, Agendas, and Voting*, Chur: Harwood Academic Publishers.

Miller, N.R. (1996), 'Information, individual errors, and collective performance: empirical evidence of the Condorcet jury theorem', *Group Decision and Negotiation*, 5, 211–28.

Miller, N.R., B. Grofman and S.L. Feld (1989), 'The geometry of majority rule', *Journal of Theoretical Politics*, 1, 379–406.

Miller, N.R., B. Grofman and S.L. Feld (1990a), 'Cycle voiding trajectories, strategic agendas, and the duality of memory and foresight: an informal exposition', *Public Choice*, 64, 265–77.

Miller, N.R., B. Grofman and S.L. Feld (1990b), 'The structure of the Banks set', *Public Choice*, 66, 243–51.

Mirkin, B. (1974), *Group Choice*, New York: Winston.

Molander, P. (1985), 'The optimal level of generosity in a selfish, uncertain environment', *Journal of Conflict Resolution*, 29, 611–18.

Moore, J. and R. Repullo (1988), 'Subgame perfect implementation', *Econometrica*, 56, 1191–220.

Morrow, J.D. (1994), *Game Theory for Political Scientists*, Princeton: Princeton University Press.

Moser, P. (1996), 'European parliament as a conditional agenda setter: what are the conditions? A critique of Tsebelis (1994)', *American Political Science Review*, 90, 834–38.

Moser, P. (1997), 'A theory of the conditional influence of the European parliament in the cooperation procedure', *Public Choice*, 91, 333–50.

Moulin, H. (1979), 'Dominance solvable voting schemes', *Econometrica*, 47, 1337–51.

Moulin, H. (1983), *The Strategy of Social Choice*, Amsterdam: North-Holland.

Moulin, H. (1986), 'Choosing from a tournament', *Social Choice and Welfare*, 3, 271–91.

Moulin, H. (1988), 'Condorcet's principle implies the no show paradox', *Journal of Economic Theory*, 45, 53–64.

Munier, B.R. (ed.) (1988), *Risk, Decision and Rationality*, Dordrecht: D. Reidel.

Munier, B.R. and M.F. Shakun (eds) (1988), *Compromise, Negotiation and Group Decision*, Dordrecht: D. Reidel.

Myerson, R. (1978), 'Refinements of the Nash equilibrium concept', *International Journal of Game Theory*, 7, 73–80.

Myerson, R.B. (1991), *Game Theory: Analysis of Conflict*, Cambridge, MA: Harvard University Press.

Nakamura, K. (1979), 'The vetoers in a simple game with ordinal preferences', *International Journal of Game Theory*, **9**, 55–61.

Nanson, E.J. (1883), 'Methods of elections', *Transactions and Proceedings of the Royal Society of Victoria*, **Art. XIX**, 197–240.

Nash, J. (1950), 'Equilibrium points in n-person games', *Proceedings of the National Academy of Sciences*, **36**, 48–9.

Nohlen, D. (1978), *Wahlsysteme der Welt*, München: Piper.

Nozick, R. (1974), *Anarchy, State and Utopia*, Oxford: Blackwell.

Nurmi, H. (1984a), 'Social choice theory and democracy: a comparison of two recent views', *European Journal of Political Research*, **12**, 325–33.

Nurmi, H. (1984b), 'On taking preferences seriously', in D. Anckar and E. Berndtson (eds), *Essays on Democratic Theory*, Helsinki: The Finnish Political Science Association.

Nurmi, H. (1986), 'Mathematical models of elections and their relevance for institutional design', *Electoral Studies*, **5**, 167–81.

Nurmi, H. (1987), *Comparing Voting Systems*, Dordrecht: D. Reidel.

Nurmi, H. (1988a), 'Banks, Borda and Copeland: a comparison of some solution concepts in finite voting games', in D. Sainsbury (ed.), *Democracy, State and Justice*, Stockholm: Almqvist & Wiksell International.

Nurmi, H. (1988b), 'Discrepancies in the outcomes resulting from different voting schemes', *Theory and Decision*, **25**, 193–208.

Nurmi, H. (1990), 'Probability models in constitutional choice', *European Journal of Political Economy*, **6**, 107–17.

Nurmi, H. (1991), 'Preferences, choices, tournaments: alternative foundations for the evaluation of voting schemes', *Quality & Quantity*, **25**, 393–405.

Nurmi, H. (1995), 'On the difficulty of making social choices', *Theory and Decision*, **38**, 99–119.

Nurmi, H. (1997a), 'Referendum design: an exercise in applied social choice theory', *Scandinavian Political Studies*, **20**, 33–52.

Nurmi, H. (1997b), 'Voting paradoxes and referenda', *Social Choice and Welfare*, forthcoming.

Nurmi, H. (1997c), 'The representation of voter groups in the European Parliament: a Penrose-Banzhaf index analysis', *Electoral Studies*, **16**, 317–27.

Nurmi, H. , T. Meskanen and A. Pajala (1997), 'Calculus of consent in the EU council of ministers', in M.J. Holler and G. Owen (eds), *Power Indices and Coalition Formation*, forthcoming.

Olson, M. (1965), *The Logic of Collective Action*, Cambridge, MA: Harvard University Press.

Ostrogorski, M. (1903), *La Démocratie et l'Organisation des Partis Politiques*, Paris: Calmann-Levy (2 vols).

Ostrom, E. (1990), *Governing the Commons: The Evolution of Institutions of Collective Action*, Cambridge: Cambridge University Press.

Owen, G. (1977), 'Values of games with a priori unions', in R. Henn and O. Moeschlin (eds), *Mathematical Economics and Game Theory*, Berlin: Springer-Verlag.

Owen, G. (1982), *Game Theory*, 2nd ed, New York: Academic Press.

Peleg, B. (1978), 'Representation of simple games by social choice functions', *International Journal of Game Theory*, **7**, 81–94.

Peleg, B. (1984), *Game-Theoretic Analysis of Voting in Committees*, Cambridge: Cambridge University Press.

Penrose, L.S. (1946), 'The elementary statistics of majority voting', *Journal of the Royal Statistical Society*, **109**, 53–7.

Pliny (1969), *Letters, Vol. 2*, Cambridge: Cambridge University Press.

Plott, C.R. (1967), 'A notion of equilibrium and its possibility under majority rule', *American Economic Review*, **57**, 788–806.

Plott, C.R. (1976), 'Axiomatic social choice theory: an overview and interpretation', *American Journal of Political Science*, **20**, 511–96.

Quattrone, G.A. and A. Tversky (1988), 'Contrasting rational and psychological analyses of political choice', *American Political Science Review*, **82**, 719–36.

Rae, D. (1967), *The Political Consequences of Electoral Laws*, New Haven: Yale University Press.

Rae, D. (1969), 'Decision rules and individual values in constitutional choice', *American Political Science Review*, **63**, 40–56.

Rae, D. and H. Daudt (1976), 'The Ostrogorski paradox: a peculiarity of compound majority decision', *European Journal of Political Research*, **4**, 391–8.

Raiffa, H. (1968), *Decision Analysis*, Reading, MA: Addison-Wesley.

Rapoport, A. and A.M. Chammah (1965), *Prisoner's Dilemma. A Study in Conflict and Cooperation*, Ann Arbor: University of Michigan Press.

Rapoport, A. and M. Guyer (1966), 'A taxonomy of 2×2 games', *General Systems: Yearbook of the Society for General Systems Research*, **11**, 203–14.

Rasmusen, E. (1989), *Games and Information: An Introduction to Game Theory*, Oxford: Basil Blackwell.

Rawls, J. (1971), *A Theory of Justice*, Oxford: Oxford University Press.

Richelson, J.T. (1979), 'A comparative analysis of social choice functions I, II,III: a summary', *Behavioral Science*, **24**, 355.

Riker, W.H. (1982), *Liberalism against Populism: A Confrontation between the Theory of Democracy and the Theory Social Choice*, San Francisco: Freeman.

Riker, W.H. (1992), 'The justification of bicameralism', *International Political Science Review*, **13**, 101–16.

Riker, W.H. and P.C. Ordeshook (1973), *An Introduction to Positive Political Theory*, Englewood Cliffs: Prentice-Hall.

Roberts, F.S. (1979), *Measurement Theory with Applications to Decision-Making, Utility and the Social Sciences*, Reading, MA: Addison-Wesley.

Roth, A. (1977), 'The Shapley value as a Von Neumann-Morgenstern utility', *Econometrica*, **45**, 657–64.

Roth, A.E. (ed.) (1988), *The Shapley Value: Essays in Honor of Lloyd S. Shapley*, Cambridge: Cambridge University Press.

Rubinstein, A. (1982), 'Perfect equilibrium in a bargaining model', *Econometrica*, **50**, 97–109.

Rubinstein, A. (1986), 'Finite automata play the repeated prisoner's dilemma', *Journal of Economic Theory*, **39**, 83–96.

Saari, D.G. (1995), *Basic Geometry of Voting*, Berlin–Heidelberg–New York: Springer-Verlag.

Saari, D.G. (1997), 'The generic existence of a core for q-rules', *Economic Theory*, **9**, 219–60.

Saari, D.G. and C.P. Simon (1977), 'Singularity theory of utility mappings - 1', *Journal of Mathematical Economics*, **4**, 217–51.

Salonen, H. and M. Wiberg (1987), 'Reputation pays: game theory as a tool for analyzing political profit from credibility', *Scandinavian Political Studies*, **10**, 151–70.

Sartori, G. (1994), *Comparative Constitutional Engineering. An Inquiry into Structures, Incentives and Outcomes*, Houndmills and London: Macmillan Press.

Satterthwaite, M.A. (1975), 'Strategy-proofness and Arrow's conditions', *Journal of Economic Theory*, **10**, 187–217.

Savage, L.J. (1954), *Foundations of Statistics*, New York: Wiley.

Schelling, Th. (1978), *Micromotives and Macrobehavior*, New York: Norton.

Schick, F. (1984), *Having Reasons. An Essay on Rationality and Sociality*, Princeton: Princeton University Press.

Schofield, N. (1978), 'Instability of simple dynamic games', *Review of Economic Studies*, **45**, 575–94.

Schofield, N. (1980), 'Generic properties of simple Bergson–Samuelson welfare functions', *Journal of Mathematical Economics*, **7**, 175–92.

Schofield, N. (1983), 'Generic instability of majority rule', *Review of Economic Studies*, **50**, 695–705.

Schofield, N. (1984a), 'Classification theorem for smooth social choice on a manifold', *Social Choice and Welfare*, **1**, 187–210.

Schofield, N. (1984b), 'Social equilibrium and cycles on compact sets', *Journal of Economic Theory*, **33**, 59–71.

Schwartz, Th. (1986), *The Logic of Collective Choice*, New York: Columbia University Press.

Selten, R. (1965), ' Spieltheoretische Behandlung eines Oligopolmodels mit Nachfrageträgheit: I–II', *Zeitschrift für die gesamte Staatswissenschaft*, **121**, 301–24, 667–89.

Selten, R. (1975), 'Reexamination of the perfectness concept for equilibrium points in extensive games', *International Journal of Game Theory*, **4**, 25–55.

Selten, R. (1978), 'The chain-store paradox', *Theory and Decision*, **9**, 127–59.

Sen, A.K. (1967), 'Isolation, assurance and the social rate of discount', *Quarterly Journal of Economics*, **81**, 112–24.

Sen, A.K. (1970), *Collective Choice and Social Welfare*, San Francisco: Holden-Day.

Sertel, M.R. and B. Yilmaz (1997), 'The majoritarian compromise is subgame-perfect implementable', *Social Choice and Welfare*, forthcoming.

Shafir, E.B., D.N. Osherson and E.E. Smith (1989), 'An advantage model of choice', *Journal of Behavioral Decision Making*, **2**, 1–23.

Shafir, E.B., D.N. Osherson and E.E. Smith (1990), 'Comparative choice and the advantage model', in K. Borcherding, O. Larichev and D. Messick (eds), *Contemporary Issues in Decision Making*, Amsterdam: North-Holland.

Shapley, L.S. and M. Shubik (1954), 'A method for evaluating the distribution of power in a committee system', *American Political Science Review*, **48**, 787–92.

Shepsle, K.A. (1974), 'Theories of Collective Choice', in C.P. Cotter (ed.), *Political Science Annual, Volume Five*, Indianapolis: Bobbs-Merrill.

Shepsle, K.A. and B.R. Weingast (1981), 'Structure-induced equilibrium and legislative choice', *Public Choice*, **37**, 503–19.

Shepsle, K.A. and B.R. Weingast (1984), 'Uncovered sets and sophisticated voting outcomes with implications for agenda institutions', *American Journal of Political Science*, **28**, 49–74.

Shugart, M.S. and J.M. Carey (1992), *Presidents and Assemblies: Constitutional Design and Electoral Dynamics*, Cambridge: Cambridge University Press.

Simon, H.A. (1957), *Models of Man*, New York: Wiley.

Simon, H.A. (1972), 'Theories of bounded rationality', in C.B. McGuire and R. Radner (eds), *Decision and Organization*, Amsterdam: North-Holland.

Simon, H.A. (1977), *Models of Discovery*, Dordrecht: D. Reidel.

Simonson, I. (1989), 'Choice based on reasons: the case of attraction and compromise effects', *Journal of Consumer Research*, 16, 158–74.

Smale, S. (1980), 'The prisoner's dilemma and dynamical systems associated to non-cooperative games', *Econometrica*, 48, 1617–34.

Srivastava, S. and M.A. Trick (1996), 'Sophisticated voting rules: the case of two tournaments', *Social Choice and Welfare*, 13, 275–89.

Steunenberg, B. (1994), 'Decision making under different institutional arrangements: legislation by the European Community', *Journal of Theoretical and Institutional Economics*, 150, 642–69.

Steunenberg, B. (1996), 'Agent discretion, regulatory policymaking, and different institutional arrangements', *Public Choice*, 86, 309–39.

Straffin, P.D. (1978), 'Probability models for power indices', in P.C. Ordeshook (ed.), *Game Theory and Political Science*, New York: New York University Press.

Straffin, P.D. (1980), *Topics in the Theory of Voting*, Boston: Birkhäuser.

Straffin, P.D. (1988), 'The Shapley–Shubik and Banzhaf power indices as probabilities', in A. Roth (ed.), *The Shapley Value: Essays in Honor of Lloyd S. Shapley*, Cambridge: Cambridge University Press.

Sugden, R. (1986), *The Economics of Rights*, Oxford: Blackwell.

Taagepera, R. and M.S. Shugart (1989), *Seats and Votes: The Effects and Determinants of Electoral Systems*, New Haven: Yale University Press.

Taylor, A. and W. Zwicker (1997), 'Interval measures of power', *Mathematical Social Sciences*, 33, 23–74.

Taylor, M. (1969), ''Proof of a theorem on majority rule', *Behavioral Science*, 14, 228–31.

Taylor, M. (1976), *Anarchy & Cooperation*, New York: Wiley.

Taylor, M. (1987), *The Possibility of Cooperation*, Cambridge: Cambridge University Press.

Taylor, M. and H. Ward (1982), 'Chickens, whales and lumpy goods', *Political Studies*, 30, 350–70.

Tsebelis, G. (1989), 'The abuse of probability in political analysis: the Robinson Crusoe fallacy', *American Political Science Review*, 83, 77–91.

Tsebelis, G. (1990), 'Are sanctions effective: a game-theoretic analysis', *Journal of Conflict Resolution*, 34, 3–28.

Tsebelis, G. (1991), 'The effect of fines on regulated industries', *Journal of Theoretical Politics*, 3, 81–101.

Tsebelis, G. (1994), 'The power of the European parliament as a conditional agenda setter', *American Political Science Review*, **88**, 128–42.

Tsebelis, G. (1996), 'More on the European parliament as a conditional agenda setter', *American Political Science Review*, **90**, 839–44.

Tsebelis, G. and G. Garrett (1996), 'Agenda setting power, power indices, and decision making in the European Union', *International Review of Law and Economics*, **16**, 345–61.

Turnovec, F. (1994), 'Voting, power and voting power', GERGE & EI Discussion Paper, Prague.

Turnovec, F. (1996), 'Weights and votes in European Union: extension and institutional reform', *European Economic Review*, **38**, 1153-70.

Turnovec, F. (1997), 'Monotonicity of power indices', prepared for presentation at the 13th International Conference on Multiple Criteria Decision Making, Cape Town, South Africa, 6–10 January, 1997.

Ullman-Margalit, E. (1977), *The Emergence of Norms*, Oxford: Oxford University Press.

van Damme, E. (1983), *Refinements of the Nash Equilibrium Concept*, Berlin: Springer-Verlag.

van Damme, E. (1984), 'A relation between perfect equilibria in extensive form games and proper equilibria in normal form games', *International Journal of Game Theory*, **13**, 1–13.

von Neumann, J. and O. Morgenstern (1944), *The Theory of Games and Economic Behavior*, Princeton: Princeton University Press.

Wagner, C. (1983), 'Anscombe's paradox and the rule of three-fourths', *Theory and Decision*, **15**, 303–8.

Wagner, C. (1984), 'Avoiding Anscombe's paradox', *Theory and Decision*, **16**, 233–8.

Westlake, M. (1994), *The Commission and the Parliament: Partners and Rivals in the European Policy-Making Process*, London: Butterworth.

Widgrén, M. (1994), 'Voting power in the EC and the consequences of two different enlargements', *European Economic Review*, **38**, 1153-70.

Wilson, R. (1985), 'Reputation in games and markets', in A. Roth (ed.), *Game-Theoretic Models of Bargaining*, Cambridge: Cambridge University Press.

Young, H.P. (1975), 'Social choice scoring functions', *SIAM Journal of Applied Mathematics*, **28**, 824–38.

Young, H.P. (1988), 'Condorcet's theory of voting', *American Political Science Review*, **82**, 1231–44.

Zermelo, E. (1913), 'Über eine Anwendung der Mengenlehre auf der Theorie des Schachspiels', in *Proceedings of the Fifth International Congress of Mathematicians*, **2**, 501–4.

Index